Applied Biomedical Microbiology

A Biofilms Approach

Applied Biomedical Microbiology

A Biofilms Approach

Edited by Daryl S. Paulson

CRC Press
Taylor & Francis Group
Boca Raton London New York

CRC Press is an imprint of the
Taylor & Francis Group, an **informa** business

CRC Press
Taylor & Francis Group
6000 Broken Sound Parkway NW, Suite 300
Boca Raton, FL 33487-2742

First issued in paperback 2017

ISBN 13: 978-1-138-11484-5 (pbk)
ISBN 13: 978-0-8493-7569-9 (hbk)

Library of Congress Cataloging-in-Publication Data

Handbook of applied biomedical microbiology : a biofilms approach / editor: Daryl S. Paulson.
    p. ; cm.
Includes bibliographical references and index.
ISBN 978-0-8493-7569-9 (hardcover : alk. paper)
    1. Biofilms--Handbooks, manuals, etc. 2. Medical microbiology--Handbooks, manuals, etc. I. Paulson, Daryl S., 1947- II. Title.
    [DNLM: 1. Biofilms. QW 90 H236 2010]

QR100.8.B55H36 2010
579'.17--dc22
                             2009025020

Visit the Taylor & Francis Web site at
http://www.taylorandfrancis.com

and the CRC Press Web site at
http://www.crcpress.com

Contents

The Editor

Daryl S. Paulson, PhD, is a decorated Vietnam combat veteran and a counselor specializing in trauma-associated disorders. He is also the president and CEO of BioScience Laboratories, Inc. He has advanced degrees in microbiology, statistics, counseling, human science, and psychology. Dr. Paulson is the author of numerous articles and books, among which are *Handbook of Regression Analysis* (Taylor & Francis, 2006), *Applied Statistical Designs for the Researcher* (Marcel Dekker, 2003), *Handbook of Topical Antimicrobials: Industrial Applications in Consumer Products and Pharmaceuticals* (editor) (Marcel Dekker, 2002), *Handbook of Topical Antimicrobial Testing and Evaluation* (Marcel Dekker, 1999), *Competitive Business, Caring Business: An Integral Perspective for the 21st Century* (Paraview Press, 2002), and *Haunted by Combat: Understanding PTSD in War Veterans Including Women, Reservists, and Those Coming Back from Iraq* (Greenwood Publishing, 2007). Address correspondence to Daryl S. Paulson, 605 Park Place, Bozeman, MT 59715, or dpaulson@biosciencelabs.com

Contributors

Judy H. Angelbeck
Consultant
Pall Life Sciences, Retired
East Hills, New York

Scott L. Burnett
ECOLAB
Eagan, Minnesota

Graham Gagnon
Water Quality and Treatment
  Centre for Water Resources
  Studies
Dalhousie University
Halifax, Nova Scotia, Canada

Darla M. Goeres
Center for Biofilm Engineering
Montana State University
Bozeman, Montana

Garth James
Center for Biofilm Engineering
Montana State University
Bozeman, Montana

Fernando M. G. Matias
Centre for Research on
  Environmental Microbiology
  (CREM)
Faculty of Medicine
University of Ottawa
Ottawa, Ontario, Canada

James W. McDowell
BioScience Laboratories, Inc.
Bozeman, Montana

John A. Mitchell
BioScience Laboratories, Inc.
Bozeman, Montana

Robert A. Monticello
ÆGIS Environments
Midland, Michigan

Daryl S. Paulson
BioScience Laboratories, Inc.
Bozeman, Montana

Marsha Pryor
Pinellas County Utilities
Clearwater, Florida

Elinor deLancey Pulcini
Center for Biofilm Engineering
Montana State University
Bozeman, Montana

Adam P. Roberts
Division of Microbial Diseases
Eastman Dental Institute
University College of London
London, United Kingdom

Syed A. Sattar
Centre for Research on
  Environmental Microbiology
  (CREM)
Faculty of Medicine
University of Ottawa
Ottawa, Ontario, Canada

V. Susan Springthorpe
Centre for Research on
  Environmental Microbiology
  (CREM)
Faculty of Medicine
University of Ottawa
Ottawa, Ontario, Canada

Kirsten M. Thompson
ECOLAB
Eagan, Minnesota

W. Curtis White
ÆGIS Environments
Midland, Michigan

chapter one

Biofilms before biofilms

W. Curtis White

Contents

Preface

Most of my life experiences are of little consequence to the topic of biofilms, but this preface provides a brief view of my experiences and how they led me to the topic of the chapter, inhibition of foundation colonization of biofilm by surface modification with organofunctional silanes, coauthored with my colleague Dr. Bob Monticello. Maybe my whole life was about biofilms?

The pathway

As we walk through life we encounter biofilms at all turns. The slippery rock coated on the shore or in the stream with communities of organisms bound in high water content matrices or the "slime" in the sink trap that plugs things up. I've always loved the image of a young person walking through the stream, slipping, and coming face down onto the slime-covered rocks. Slightly dazed, his or her vision focuses onto the wet and mysterious association of algae, emergents, and suspected other creatures and plants in the biofilm he or she has slid onto in such an ungracious way. Like Horton carrying around the dandelion seed head, a world is disclosed, or like Alice, slipping through the rabbit hole, life is magnified in this newly encountered and recognized world of biofilms. We think little of these biofilm communities until we put our science hats on and peer into this fascinating world. And, equally important, that encounter with a biofilm did not exist until it was delved into, a communication vocabulary was proposed and developed, the story was told, and peers verified the observations (maybe without the slip and fall). How elegant, the scientific method actually works, and a sliver of science marches on.

The simplistic view of conditions for microbial growth and biofilm existence is nutrients, water, proper temperatures, and the need for receptive surfaces. Then all of the complexities of matrix and community building and destroying begin. Our pathway arms us with some insights, and then we step back and make those new hypotheses that grow our knowledge and lead to a new plateau of discovery and definition. We pass like Alice, from her tiny person stage and view of the world, to the giant view as she expands her view and we see, mediated by good science and discipline and not by magic mushrooms, a biofilm microcosm set down in and influenced by greater forces and elements of energy and matter, all around its microhabitat.

For me, these simple biofilms are great cities with all of the infrastructures and complexities of a community—highways, hospitals, food stores, waste systems, emergency services, schools, life, death, and energy plants, all found in a matrix on a surface. With a modern twist and with apologies, I think of these as "Coster-towns." For all of you interested in biofilms, this is a great and important adventure. They impact our lives for good and bad and hold the keys to the foods and energy of the future.

I have been an industrial microbiologist for 46 years and find that in my work with water-based product preservatives, recirculating water biocides, textile preservatives, wound care products, bucal and vaginal antimicrobial products, disinfectants/sanitizers, wood preservatives, paint film preservatives, marine antifoulants, and other substrate preservatives, my colleagues and I were constantly dealing with biofilms and the positive and negative things that they can do. On the medical side, even Gram, as he studied tissue stains and saw the importance to microbiology, knew there were communities at work and dynamics at the tissue-surface microbe interfaces.

Learning by teaching

During my teaching years there were a series of elements that provided a basis for very important learning that relates to biofilms. The years when I was teaching, the mid-1960s to the early 1970s, were filled with environmental concepts, and teaching was heavily based on ecosystems and biospheres. It's easy to see how biofilms fit into this scheme. In biology there are several basic principles that are used to describe life and life processes. Keys for biofilms are:

- Succession
  1. Settle, grow, reproduce, die
  2. Change the environment that favors species 1, then 2, then 3, etc.
  3. Aerobic to microaereophilic to anaerobic
  4. Nutrient sources and changes
  5. Waste control and changes
  6. pH changes
  7. Temperature changes
  8. Osmotic stress changes (salts and water availability)
  9. Active and passive exchange
- Reproduction
  1. Genetic survival and change
  2. Environmental and social stress
     a. Biofilms provide microcosms of stress. This forces all types of reproduction.
     b. By organization: The city of Calyx (Coster-town), with buildings, rooms, and cages and dynamic transport between them all, allows for the proximity and transfer routes critical to a dynamic biofilm.
- Conservation of matter and energy
  1. This drives individuals and systems, but in ecosystems (biofilms or leaf litter) where the individuals have biochemical mechanisms for balance—homeostasis and entropy to enthalpy operate—the equation of catabolism and anabolism plays itself out in a dynamic but stable way.
  2. Biofilms provide for the dynamics of efficient use of energy and matter.

Solving problems in the real world

Our goal in an industrial setting was to take advantage of what we understood about biofilms to control them as best as possible. This was not to create a taxonomy or nomenclature or to study the minutia of these often destructive associations of microorganisms, but to eliminate them, or at least control them, within the needs of a specific industrial process or in products. At that time in the early 1960s we were building on the knowledge brought forth by those myriad investigators that studied microbial communities (biofilms) in soil and root zone science and those that were engineering biodigestion and bioreactor systems. That knowledge has exploded in the last 20 years and challenges investigators with looking beyond the elements they study and stepping back to look at the whole systems of living and inanimate objects and materials.

Biofilms, when defined as associations of microbes in close proximity to each other within a matrix on surfaces, allow us to understand these matrices with all of the complexity of their substrates, the binding or separating nature of the matrix, whether in layers, swirls, or continuous or discontinuous phases (think as if they might be emulsions or invert emulsions), varied oxygen tensions, availability of nutrients, ways of getting rid of wastes, and pH profiles all intertwined with the problems and opportunities associated with microbial metabolism and reproduction.

Building on the knowledge gained from the soil scientists and rhizome microbiologists, it was easy to relate the biofilms in root masses and their functions and dynamics to the problems we were trying to deal with caused by the biofilm slimes with their odors, corrosion, and deterioration capabilities in recirculating water systems, paint or other water-based products packaging, paint film surfaces at the substrate interface and environmental interface, human body ecology, and water transmission systems.

It is interesting to note the importance of surfaces to these biofilm associations and to the problems that they cause. What occurred in one root mass, on wood, steel, mucous membranes, or roofing tiles was far more significant than the basic biological phenomenon of succession that played out as a contaminated surface aged. It was clear that the nature of the substrates, chemically, surface energy-wise, and topography, were all influencing the base layer of microbial colonization and the subsequent successional activities of the microflora.

In those early years at Dow Chemical, after working on the bio-leaching of molybdenum and tellurium from their ores, I got involved with the workhorse of all biofilms when we were building substrates (Surf-Paks) for sewage disposal plants. The challenge was to maximize flow of sewage over the so-called zoogleal mass to maximize conversion and reduction of bioloads and other contaminants. The physics and mathematical descriptions with continuous and discontinuous (turbulent) phases drove an integration of biochemical conversion and contact methods with the time needed to facilitate and maximize the reduction of the wastes. Even in free-flow or falling film digesters, the biofilm communities on stationary substrates and all of their phases of anaerobic, microaereophilic, moderate oxygen tension, and aerobic conditions, their potential for nutrient and waste product movement, and all of the physical forces against the substrate and within the biomass drove the digestion dynamics. For example, in such digesters or fermentation vessels the addition of fumed silica ($500 + m^2/g$) efficiencies went up compared to pure cellular and fluid systems. Providing flow agents, chemical or physical, also increased efficiency.

Another perspective on biofilms before biofilms is the curious community of microorganisms that form at oil-water interfaces. These associations dominated by iron bacteria or sulfur bacteria can be copious slime producers. My experience with these rests in one of my earliest projects in the Dow Chemical laboratories in 1963 with aviation fuels. In the late 1950s and early 1960s, there were some airplane crashes due to slimes (floating biofilm slimes) from fuels plugging the fuel injectors of the turbo jets. Our laboratory was mobilized to find a solution to this problem. We worked with our full arsenal of antimicrobial agents. One fundamental of this was that microbes live in the water phase (where condensation in the large fuel storage tanks allowed for the interface) and take advantage of soluble and interface nutrients and the ability to move waste materials from the water to the oil phase. So we wanted to deliver the antimicrobial in the oil phase but have it partition into the water phase. We had some great chemistry for this, but the best answer to the slimes came from some fundamental understanding of the "floating" biofilms at the oil-water interface. A simple increase of the concentration of the glycols used as antifreeze in the fuel changed the environment, and the iron bacteria and their slimes were eliminated. These were not the typical slime (capsular material)-producing organisms, such as the "rope yeasts" or even *Pseudomonas aeruginosa* PRD-10 strain types of organisms. This was truly a dynamic community of a biofilm, but here the surface interfaces were oil-water surfaces rather than air-solid surfaces. These types of interfaces are abundant in nature and in human-made products and structures.

I also was taken, as a 19-year-old co-op, into the brewing industry and into a world of biofilms at work. Wow, free beer and another reason to want to be a microbiologist. In beer processing, the bottled beer is pasteurized in a series of sequential water spray booths where the needed pasteurization temperatures and residence times can occur without the temperature shocks that would ruin flavor and cause bottles to explode. Well, some do break, and that bath of nutrients, with the enviably positive temperatures, all of the water, the ready availability of yeast and other microbial organisms, and the stainless steel pasteurizer equipment surfaces, set in play a chance for biofilms. And, what biofilms they were. Thick slabs, up to 2 in., that were continuous and strong enough such that you could lift the "mat" and shake it, rippling out a gelatin-like wave. Think of the varied conditions throughout that biofilm and the interfaces at the metal and the air-water. Odors, for sure, from the anaerobic hydrogen sulfide to the aerobic yeasty odors; corrosion, for sure, where even the stainless steel was pitted by our sulfate reducers and their waste products; and the slimes mass that plugged up the equipment and the process. The economic reality of this problem causing biofilm was to stop it from starting, which we did by population control.

Again drawing on my experiences and the mentoring of Dr. Dorthea Mangum at Dow Chemical, surfaces and the nature of the foundation organisms were critical to the control of marine biofouling. Doing work on marine antifouling and the colonization of surfaces with *Enteromorpha* sp. algae or *Balanus* sp. barnacles (both with one-celled stages and identical chemical attachment mechanisms), the importance of biofilms is reinforced at every turn. Dorthea introduced me to the idea that divalent cations (Mg, Mn, etc.) associated with the surfaces from the seawater or the substrate and attracted or facilitated the attachment of foundation bacteria or fungi, and the subsequent attraction and attachment of the cyprids and algal precursor cells. Her work and observations were done in the seawater canals and concentration areas where Dow was extracting magnesium from seawater at Freeport, Texas. Our tactics for control were to alter the surfaces with chelating agents and neutralizing coatings with slow-release antibacterial, antifungal, or antialgal compounds, and by altering surface energies. I wish that we had been able to publish on some of this work done in the early to mid-1970s. One thing we learned was that there was a succession of colonization where the bacterial and fungal communities established because divalent ion concentrations on the metal test coupons were important to the later colonization of the algae, barnacles, and bryozoans. Thus, if we controlled the early colonizers, we could alter the rate and durability of the higher organisms that caused the slowing of the ships and the corrosion cells on the metal hulls.

Later in my career I got heavily involved in the "sick building syndrome" events, and with my experiences in preservation of building materials, I came into the control of microbiological influencers with a very strong background. This had me climbing around in crawl spaces, above ceiling spaces, in air handling systems, and into architectural drawings to understand the myriad microbiological niches, the problems that they cause, and the tactics we would use for control. Minimizing microbiological problems in these settings takes a good understanding of the lifestyles of the organisms, the modes of action and delivery mechanisms of your control technologies, and the target environmental niches. This leads us back to biofilms. The biofilm associations in the cooling towers have been well studied because of the fascinating association of the health impacting *Legionella pneumophila*. Here, after we learned to culture and identify the causative organism, we learned all kinds of subtle and important things about this organism in the biofilm. Without the recognition of the biofilm association we would not have nearly approached practical control strategies. The microbial associations from cooling towers and their heat exchangers, above ceiling spaces, wallboards, ceiling tiles, water cooler areas, condensation on water supply pipes, fire control system pipes, window or other fenestration areas, to carpeting and more, are all biofilm communities at one or another phase of their growth. We are certainly surrounded by biofilms.

The human body

Now for our medical friends, the world of biofilms—and their importance to health and disease—is just beginning to be understood and tapped for its benefits. Certainly some of the most fascinating biofilms are those that exist in the human body. I've had the opportunity to work on mouthwashes and astringents, catheter antimicrobial treatments, and antimicrobial treatments of other implantables and insertables. Study of the biofilm communities in the bucal cavity have led to some wonderful techniques and some unique knowledge of biofilms in the human body. But this is only a small part of the story. Biofilms found on a catheter, synthetic heart valve, or other insertable or implantable devices, or as part of the plaques in veins, arteries, spinal tissues, or joints, wreak havoc in the body, causing untold amounts of morbidity and mortality. Understanding the importance of microbial flora on the surfaces of the digestive system for the process of digestion seems like an easy thing to accept and deal with, but the dynamics of this giant biofilm reactor are actually very complicated. What we have learned about ulcers and the pockets of *Clostridium difficili* in the alimentary canal is undoubtedly the tip of the iceberg of knowledge about these biofilms. While working with contact lens problems, the interface at the mucus membrane, the individual immune system, and the associated bioflora, the dynamics produced a series of problems with the lens. This medical world of microorganisms can be further expanded by viewing the microbiological problems associated with medical garments and other woven and nonwoven materials. Again, biofilms are causing health and economic problems.

For all of these areas of biofilm understanding, the emerging capabilities to understand the eco-dynamics of nonculturable microorganisms open up new ways of thinking about biofilms. In human health we know that vaginal yeasts at the labia or in mucosal tissue can produce trigger chemicals that excite the immune system and the nervous system to cause a variety of discomforting reactions to infection. Many of the trigger chemicals are suspected of coming from nonculturable microorganisms. At every turn in the world of biofilms, there are new tools to study, new challenges as we learn more, and more opportunities.

Surfaces and biofilm control

This history of experiences led me to some basic understandings of microbial life, alteration, and propagation at surface interfaces. I had worked with altering surface energies and learned that the lowest surface energy was not the best to stop biofilms or to allow for their easy removal. I learned that ablative coatings had limited success. I learned that use of slow-release toxins was almost impossible to control in terms of toxicity to nontarget species, adaptation, and straight-out effectiveness in terms of dose-response. This, in turn, led to the realization that microbial control tactics might best be focused on the surfaces or at the interfaces where foundation organisms establish residency. What was left was to fully explore the concept of having a surface that is inherently antimicrobial in an active sense but with a passive mode of action. That is the subject of our Chapter 4 and has proved to be an intriguing way to deal with microbial colonization and propagation on surfaces.

The twilight leads to dawn

The reality of all of this is that microorganisms do not exist without surfaces, and hence all of their positive and negative effects are manifested as biofilms. Yes, they are transient in phases, floating in the air, water, or other media, but life is really manifested and transformed on surfaces in the always intriguing world of biofilms.

The exciting part is that biofilms are everywhere and that recognition allows us, demands of us, to look at the community and successional nature of these associations and to always take the most holistic view possible. Unquestionably, this breadth of view allows for both the needed convergent and divergent thinking that will tap into the potential of biofilms for the good that they do, and allow us better control of the bad that they do, and truly give us the keys to Coster-town.

For me, I have had the pleasure and good fortune to work with an incredible array of dedicated scientists where they have mentored me and allowed me to fail. In my analogy of the community, I have been in the amusement park and hope to stay there for a good while more, and that those who read this will add to the fun of my ride with new insights into the world, the universe, of biofilms.

chapter two

Efficacy of preoperative antimicrobial skin preparation solutions on biofilm bacteria*

Daryl S. Paulson

Contents

Until recently, perioperative professionals were taught that microbial infections (e.g., bacterial, viral, fungal) were caused by free-moving, individual microorganisms or small isolated groups of microorganisms. Microorganisms causing infections entered the body via a wound or by direct invasion, spread through the body, multiplied in the body, evaded immunological defenses (i.e., T lymphocyte, B lymphocyte, and phagocytic activity), and were shed from the body to infect new hosts [1].

Although it is true that in acute infections, bacteria generally are found in a free-floating (i.e., planktonic) form, if bacteria (i.e., prokaryotes) establish a presence of any duration in the body, they generally form a highly complex, self-regulating, bacterial community known as a biofilm matrix [2]. Bacterial biofilm complexes cause challenging infections that

* Reprinted from Daryl S. Paulson, "Efficacy of Preoperative Antimicrobial Skin Preparation Solutions on Biofilm Bacteria," AORN 81 (2005): 491–501. Copyright 2005, with permission from Elsevier.

generally are both difficult and expensive to treat and manage. Terms relevant to biofilm bacteria are defined in Table 2.1.

Through a process termed quorum sensing, bacteria in a biofilm matrix can chemically communicate system-level needs for the well-being of the entire biofilm community. Quorum sensing between bacteria enables a biofilm community to induce or repress specific gene expressions, regulating such activities as cell division, metabolic rates, production of virulence factors, plasmid transfer for antibiotic resistance, and release of planktonic bacteria from the biofilm [3–5].

Biofilm infections often begin during surgical procedures, such as insertion of vascular catheter lines, pacemakers, heart valves, permanent biomaterials for repair of aneurysms, or prosthetic joint replacements. Other medical procedures associated with biofilm establishment are intratracheal intubation needed for ventilators and protracted use of indwelling urinary catheters [6]. It is important for perioperative staff members to recognize that implantation of devices or biomaterials may lead to the formation of biofilms, which increases the risk of difficult-to-treat infections in postoperative patients.

Table 2.1 Definitions

Antimicrobial residual properties:	When chlorhexidine gluconate or zinc pyrithione solution is used for at least 2 to 3 days before surgery as a presurgical site wash, the medications are adsorbed into the stratum corneum and prevent normal microbial population regrowth. Then, when the preoperative skin preparation is performed just before a surgical procedure, the normal microbial population numbers already have been reduced greatly, and the preoperative skin preparation is more effective.
Biofilm:	A complex community of microorganisms enclosed in an exopolysaccharide matrix attached to tissue or an inanimate surface.
Coadhesion:	The process of planktonic microorganisms binding to microorganisms attached to surfaces.
Exopolysaccharides:	Polymerized material produced by microorganisms that constitutes the biofilm, providing protection and containment for the microorganisms.
Laminin:	Linking proteins of basal lamina, which induce adhesion and enhance spreading of microorganisms in a biofilm.
Planktonic:	Free-floating microorganisms.
Quorum sensing:	Chemical communication between microorganisms.
Van der Waal's force:	Nonspecific attraction between atoms that are 3 to 4 angstrom units (Å) apart.

Biofilm genesis

To form a clinically significant biofilm, bacteria must attach to tissue or an inanimate surface (e.g., titanium, stainless steel, polytetrafluoroethylene, polyester fiber) in a patient's body and then attract and attach to other bacterial cells [47]. Typically, direct attachment of bacteria to tissue elicits such a strong immunological response (e.g., high fever, malaise) that it becomes apparent, so patients are treated immediately, according to standard protocols, before a biofilm is able to develop. In comparison, implanted biomaterials, prosthetics, and devices have inanimate surfaces. Bacteria adhering to these inanimate surfaces do not elicit an immune response. The lack of an immune response results in patients not being treated for infection; thus, normal skin bacterial residents (e.g., *Staphylococcus epidermidis*) attaching to implanted materials can lead to the development of a biofilm.

Bacterial attachment to inanimate surfaces generally requires that a surface be conditioned by organic deposits, such as collagen, laminin, fibrin, and fibrinogen. The bacterial cells and organic deposits are mutually attracted via noncovalent forces, including van der Waal's forces and hydrophobic interactions [7–9]. Bacteria with receptor sites for these organic compounds can attach directly to the compounds by primary adhesion (Figure 2.1), divide, and produce an exopolysaccharide biofilm matrix. This establishes a bacterial presence that is protected from the body's natural immunological surveillance, including phagocytosis and antibiotic treatments [7].

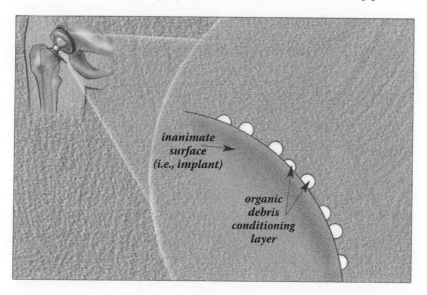

Figure 2.1 **(See color insert following page 22.)** Surface conditioning. Organic debris conditions the inanimate surface of implants and other biomaterials. (Reprinted from *AORN*, Vol. 81, No. 3, Daryl S. Paulson, PhD. Efficacy of Preoperative Antimicrobial Skin Preparation Solutions on Biofilm Bacteria, 491–501, 2005. With permission from Elsevier.)

The conditioning layer influences which organisms will be the primary colonizers of the biofilm (Figure 2.2). For example, specific organic substances (i.e., collagen, laminin, fibrinogen, fibrin) are deposited on the inanimate surface for *Staphylococcus aureus* to produce an exopolysaccharide biofilm matrix. Similarly, for *Staphylococcus epidermidis* to produce a biofilm matrix, fibrinogen-binding protein is deposited on the inanimate surface.

Biofilms also contain bacteria that are attached to other bacteria and not to organic debris (Figure 2.3). Different bacterial species that cannot attach to organic material themselves are able to attach to bacteria already adhering to the organic material. Specific and compatible attachment sites are required of both bacteria; this attachment process is called coadhesion. The exact extent of this phenomenon and its process is not well understood at this time, but much anecdotal knowledge derives from oral biofilms (i.e., plaque), for which the coadhesive process is understood more fully [10,11].

Individual bacterial cells can form an elaborate matrix of exopolysaccharide and interstitial fluid consisting of 95 to 99% water, 2% bacterial content, and 1 to 2% exopolysaccharide content [2]. A biofilm matrix may appear as depicted in Figure 2.4 or as a thin layer of bacteria in an exopolysaccharide matrix. To date, no universal configuration has been determined for medical biofilms.

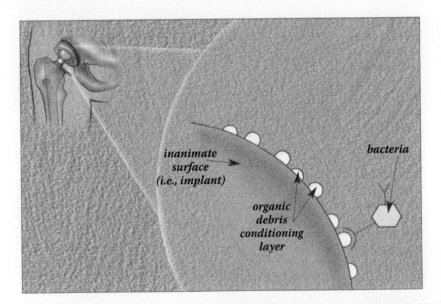

Figure 2.2 **(See color insert following page 22.)** Bacterial attachment. Bacteria can attach to the organic debris by chemical adhesion. (Reprinted from *AORN*, Vol. 81, No. 3, Daryl S. Paulson, PhD. Efficacy of Preoperative Antimicrobial Skin Preparation Solutions on Biofilm Bacteria, 491–501, 2005. With permission from Elsevier.)

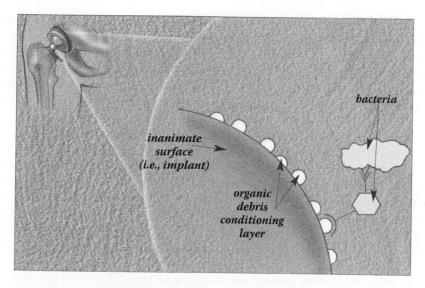

Figure 2.3 **(See color insert following page 22.)** Coadhesion. Some bacteria that cannot attach directly to the organic debris can attach to bacteria that can attach to organic debris. (Reprinted from *AORN*, Vol. 81, No. 3, Daryl S. Paulson, PhD. Efficacy of Preoperative Antimicrobial Skin Preparation Solutions on Biofilm Bacteria, 491–501, 2005. With permission from Elsevier.)

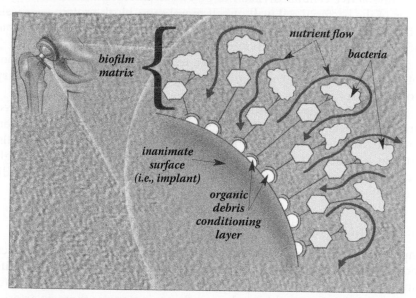

Figure 2.4 **(See color insert page following 22.)** Biofilm matrix. The individual bacterial cells form an elaborate matrix of exopolysaccharides and interstitial fluid. (Reprinted from *AORN*, Vol. 81, No. 3, Daryl S. Paulson, PhD. Efficacy of Preoperative Antimicrobial Skin Preparation Solutions on Biofilm Bacteria, 491–501, 2005. With permission from Elsevier.)

A biofilm matrix offers bacterial protection and thereby increases resistance to the immunological responses of both humoral and cellular derivation, as well as from the phagocytic activities of neutrophils and tissue macrophages [2–4]. This facilitates growth of the biofilm matrix (e.g., on a venous catheter), and allows the biofilm to slough off planktonic bacteria that can produce septic conditions throughout a patient's body. Biofilm matrices often are slow growing and localized, however, affecting only an implant and surrounding tissues, and this may require surgical removal of the implant and debridement of associated tissues [12–15].

Bacteria in a biofilm are 500 to 1,500 times more resistant to antibiotic therapy than are planktonic bacteria [16,17]. Initially, researchers believed that the exopolysaccharide matrix provided a barrier that protected the bacteria from direct exposure to antibiotics [2,3]. It now appears that the reason is more complex [16,17]. Bacteria in biofilm are more metabolically efficient, which limits their uptake of antibiotics. Although bacteria in a biofilm generally do not replicate as rapidly as they do in the planktonic state, different biofilm sections are in various stages of growth (i.e., static, stationary, exponential) at any given time [18,19]. On average, however, the entire growth rate of the biofilm community appears quiescent [15,16]. Bacteria in the biofilm that are most susceptible to antibiotics are in the exponential growth phase because antibiotics require high metabolic rates and active cellular division to be effective. Antibiotics inhibit synthesis of the bacterial cell wall or cell membrane, block protein synthesis at the 30s or 50s ribosome subunit, block DNA replication, or block folate coenzymes needed in DNA synthesis [18].

Destroying biofilm sections in the exponential growth phase does not destroy bacteria in the biofilm that are in the static and stationary growth phases.

Resistance to some disinfectants (e.g., hydrogen peroxide) is related directly to bacterial density in a biofilm. Degradation of hydrogen peroxide via catalase produced by bacterial cells, including nonviable cells, requires a concerted systems effort by a group of bacteria. A single bacterium is not able to produce enough catalase to overcome the debilitating effects of hydrogen peroxide [19,20].

Perioperative implications

Implanted devices (e.g., hemodialysis grafts, genitourinary prosthetics, pacemaker leads, prosthetic heart valves, vascular grafts) have significant potential for incurring biofilm infection [8]. For example, polyester fiber grafts that are used to replace and repair stenotic thoracic arteries and abdominal aortic aneurysms are prone to biofilm infections from coagulase-negative *Staphylococcus* species, *Staphylococcus aureus*, and other microorganisms.

Vascular catheterization

Biofilms are a serious concern for patients who have vascular catheters. Microorganisms, particularly normal skin flora, colonize and form biofilms quickly on catheter surfaces; however, contaminative exogenous microorganisms from health care personnel, contaminated infusion fluid, and distal infections transported via hematogenous routes also have been implicated [7,12,21]. Many of the millions of patients who undergo vascular catheterization procedures in the United States every year become infected via biofilms [16]. Four to 14% of patients who have a central venous catheter experience septicemia [15]. Central venous catheter infections most commonly are caused by normal skin bacteria, including *Staphylococcus epidermidis* or other catalase-negative *Staphylococcus* species. Other common bacteria cultured from these catheters include *Staphylococcus aureus*, *Pseudomonas aeruginosa*, and *Enterococcus* species [21].

Local biofilm infections, which are very common at catheter insertion sites, include tunnel infections (i.e., cellulitis along the subcutaneous catheter route) and frank catheter-tip colonization. These can lead to life-threatening septicemias.

The current trend for topical antimicrobials is to demonstrate antimicrobial persistence for longer periods of time to limit injury to veins from multiple insertions; however, patients with catheters may have an increased risk for acquiring biofilm infections when the catheter remains at a specific site for a longer time. To counteract this threat, some catheter manufacturers are partnering with manufacturers of topical antimicrobials to produce tubing, cannulas, and catheter insertion tips treated with antimicrobial products, such as silver sulfadiazine or chlorhexidine gluconate [22].

Nearly 100% of superior vena cava catheters become locally infected within 2 to 3 days of placement [16]. Many of these infections are caused by coagulase-negative *Staphylococcus* species, especially *Staphylococcus epidermidis*. A high proportion of these bacteria demonstrate resistance to multiple antibiotic medications, especially methicillin and oxacillin [16].

Devices exposed to direct blood flow, such as vascular catheters and heart valves, pose a serious risk for both local and systemic infections. It is important for perioperative nurses to understand Virchow's triad (i.e., surface area, blood contact, and flow rate), and that the greater the surface area, the more probable it is that bacteria can colonize it. Direct blood flow offers a continuous source of conditioning material that coats a device in preparation for bacterial attachment. The blood flow also exerts a shearing effect that can transport planktonic bacteria and biofilm clumps to other areas in the body [8,23]. Finally, ventricular-peritoneal shunts used to reduce intracranial pressure almost always become biofilm contaminated with *Staphylococcus epidermidis* and *Staphylococcus aureus* [15].

Orthopedics

Joint replacements (e.g., hip, knee) pose the threat of postoperative biofilm infections with particularly devastating effects, including osteomyelitis. *Staphylococcus epidermidis, Staphylococcus aureus,* and *Pseudomonas aeruginosa* commonly are cultured from these implants [13]. In many situations, biofilm-infected joints require revisional surgery with a second implant, and this can be expensive and traumatic. Special precautions with topical skin antiseptics may be valuable because *Staphylococcus epidermidis* is prevalent in these infections.

Endotracheal tubes

Patients who remain intubated after a surgical procedure are prone to ventilator-associated pneumonia. Endotracheal tubes bypass the body's normal pulmonary clearing responses (e.g., coughing, mucociliary clearance), increase mucus secretions because of irritation and inflammation, and tend to denude cilia from the tracheal epithelium. This allows secretions to enter the lungs through the stented glottis [8]. Biofilms quickly develop on the endotracheal tube and can pass easily into the lungs, particularly during suctioning procedures.

Are topical antimicrobials effective?

It is pertinent to determine the effectiveness of topical antimicrobials that are used to remove germs from the skin before catheter insertions or before preoperative skin preparation, or when challenged with bacteria in a biofilm matrix. Currently, antimicrobial efficacy testing is performed almost exclusively on bacteria in the planktonic state. An evaluation to determine efficacy against biofilms was performed using a number of common topical skin antiseptics to challenge pathogenic bacterial species prevalent in biofilm infections. The evaluation determined the resistance to killing provided by a biofilm matrix compared to the bacteriocidal effectiveness of each antiseptic versus the bacteria in a planktonic state.

Materials and methods

The bacterial species and strains used were supplied by the American Type Culture Collection. These included methicillin-resistant *Staphylococcus aureus*, methicillin-resistant *Staphylococcus epidermidis, Pseudomonas aeruginosa,* and vancomycin-resistant *Enterococcus faecium.*

The topical antimicrobial compounds evaluated are used commonly to prepare patients' skin before surgery or vascular catheter insertion. The active ingredients in these antimicrobial compounds include 70% isopropanol + 2% chlorhexidine gluconate, 72% isopropanol + 7.5% povidone iodine, 73% ethanol + 0.25% zinc pyrithione, and 62% ethanol + <5% isopropanol.

Results

The results of the evaluation are presented in Table 2.2. Planktonic time kills were performed on bacterial solutions containing approximately 1×10^9 colony forming units (CFU)/ml. A 0.1 ml aliquot of the suspension was added to 9.9 ml of product to result in a 99% concentration of the product. Exposure times were 15 s and 2 min. After each exposure, a 1 ml portion of the product/bacterial solution was transferred to 9 ml of a phosphate-buffered solution with product neutralizers. It was serially diluted, plated, and incubated at $35 \pm 2°C$ ($95 \pm 3.6°$ F).

Biofilms developed on microporous membranes resting on agar nutrient medium. The membranes were inoculated with a cell suspension containing approximately 1×10^9 CFU/ml to generate the biofilms.

The membrane-supported biofilms were incubated at $35 \pm 2°C$ ($95 \pm 3.6°F$) for 48 h, with transfers to fresh agar nutrient medium approximately every 10 to 12 h. After approximately 48 h, the membrane-supported biofilms were exposed to each antimicrobial product in screw-capped containers for 15 s and 2 min. Neutralizing fluid was added to the jars after each designated exposure time, and a vortexing unit was then used to agitate the contents to disaggregate the biofilm. The neutralizer/product/disaggregated biofilm suspension then was serially diluted, plated, and incubated at $35 \pm 2°C$ ($95 \pm 3.6°F$).

For each antimicrobial compound tested, with the exception of *Enterococcus faecium*, biofilm-enclosed bacteria retarded the microbial action of alcohol, alcohol and povidone iodine, alcohol and chlorhexidine gluconate, and alcohol and zinc pyrithione, relative to the planktonic time kill. Generally, however, the topical antimicrobials tested demonstrated high antimicrobial activity against the biofilms within a time frame practical for site preparation. Unlike antibiotics that kill by interrupting bacterial replicative and synthesizing mechanisms, alcohols coagulate and denature bacterial proteins and leach membrane lipids; chlorhexidine gluconate punctures the cytoplasmic membrane so that low molecular weight cytoplasmic components, such as potassium, leak out; and povidone iodine oxidatively blocks disulfide bridging, which is important in bacterial protein synthesis [24–26].

Table 2.2 Time Kill Results

	Log$_{10}$ reduction from initial population			
	Planktonic		Biofilm	
	15 s[a]	2 min[a]	15 s[a]	2 min[a]
Pseudomonas aeruginosa				
62% ethanol + <5% isopropanol	>5	>5	0.35	>5
Staphylococcus aureus				
70% isopropanol + 2% chlorhexidine gluconate (CHG)	>6	>6	1.51	>6
72% isopropanol + 7.5% povidone iodine	>6	>6	0.37	>5
73% ethanol + 0.25% zinc pyrithione	>5	>5	0.10	>5
62% ethanol + < 5% isopropanol	>6	>6	0.08	—[b]
Methicillin-resistant Staphylococcus aureus				
70% isopropanol + 2% CHG	>6	>6	3.14	>6
72% isopropanol + 7.5% povidone iodine	>6	>6	0.75	>6
62% ethanol + < 5% isopropanol	>6	>6	0.08	>6
Staphylococcus epidermidis				
70% isopropanol + 2% CHG	>6	>6	1.86	>6
72% isopropanol + 7.5% povidone iodine	>6	>6	0.97	>6
73% ethanol + 0.25% zinc pyrithione	>5	>5	0.28	>5
62% ethanol + <5% isopropanol	>6	>6	0.01	>6
Methicillin-resistant *Staphylococcus epidermidis*				
72% isopropanol + 7.5% povidone iodine	>5	>5	1.70	>5
62% ethanol + < 5% isopropanol	>5	>5	0.36	>5
Vancomycin-resistant *Enterococcus faecium*				
70% isopropanol + 2% CHG	>5	>5	>5	>5

[a] Exposure time.
[b] Unable to validate because of several outliers in data.

Recommendations

An effective, persistently active antimicrobial formulation should be used to prepare skin sites thoroughly before performing vascular catheterization procedures or surgeries requiring stents, biomaterial repairs, joint replacements, or orthopedic implants. It may be prudent to have patients treat intended surgical sites with alcohol/chlorhexidine gluconate, alcohol/zinc pyrithione, or a solution containing only chlorhexidine gluconate at home for 3 or 4 consecutive days before surgery [26]. With repeated use, both chlorhexidine gluconate and zinc pyrithione demonstrate residual antimicrobial properties that prevent skin colonization from rebounding to baseline microbial population levels [27]. Approximately 3 days of repeated exposure, however, is necessary for the chlorhexidine gluconate or zinc pyrithione to be adsorbed onto the stratum corneum [28]. It is important, therefore, to use antimicrobially effective surgical scrub products labeled with both immediate and persistent antimicrobial properties [29].

Prepping with alcohol and chlorhexidine gluconate, alcohol and povidone iodine, alcohol and zinc pyrithione, or chlorhexidine gluconate alone before vascular catheterization procedures may reduce catheter-associated infections. Using antimicrobially treated bandaging may improve prospects even more. Currently, catheters coated with antimicrobials are being evaluated for their value in preventing bacterial attachment and biofilm formation [8].

As in many surgeries, it is important to administer prophylactic antibiotics just before and during the surgical procedure. It is prudent to employ antibiotics that can inactivate methicillin-resistant *Staphylococcus aureus*, methicillin-resistant *Staphylococcus epidermidis*, and vancomycin-resistant *Enterococci* [30]. The goal is to eliminate planktonic bacteria before they can form a biofilm that is resistant to antibiotics [30].

Health care practitioners must take special care for orthopedic surgeries involving implants in joint replacements. Antimicrobial incision drapes are recommended to isolate the surrounding skin surface from the incision site [31]. An alcohol skin wipe also should be performed before placement of an antimicrobial incision drape [32].

It is important to take a proactive approach to preventing contamination of wounds and implants. Health care professionals should require vendors to provide published documentation demonstrating that products are effective against microorganisms in the biofilm matrix, as well as in the planktonic state.

Acknowledgments

The author acknowledges Terri Eastman, manager of *in vitro* laboratories; John A. Mitchell, PhD, director of quality assurance; and Karen Wesenburg-Ward, PhD, project manager of biofilm division, BioScience Laboratories, Inc., Bozeman, Montana, for their assistance in producing this article. BioScience Laboratories manufactures technologies described in this article. The evaluation described in this article was funded by BioScience Laboratories, Inc. Publication of this article in no way implies AORN endorsement of products manufactured by BioScience Laboratories, Inc.

References

1. C Mims, A Nash, J Stephen, *Mim's Pathogenesis of Infectious Disease*, 5th ed. (San Diego: Academic Press, 2001).
2. M G Darby, G A O'Toole, "Microbial Biofilms: From Ecology to Molecular Genetics," *Microbiology and Molecular Biology Reviews* 64 (2000): 847–67.
3. A L Reysenbach, E Shock, "Merging Genomes with Geochemistry in Hydrothermic Ecosystems," *Science* 296 (2002): 1077–82.
4. J Jass, S Surman, J T Waller, "Microbial Biofilms in Medicine," in *Medical Biofilms: Detection, Prevention and Control*, ed. H Jass, S Surman, J Waller (West Sussex, UK: John Wiley & Sons, 2003), 1–28.
5. M Hentzer, M Givskov, L Eberl, "Quorum Sensing in Biofilms: Gossip in Slime City," in *Microbial Biofilms*, ed. M Ghannoum, G A O'Toole (Washington, DC: American Society for Microbiology, 2004) 118–40.
6. S N Wai, Y Mizunoe, J Jass, "Biofilm-Related Infections on Tissue Surfaces," in *Medical Biofilms: Detection, Prevention and Control*, ed. J Jass, S Surman, J Waller (West Sussex, UK: John Wiley & Sons, 2003), 1–28.
7. F Gotz, G Peterson, "Colonization of Medical Devices by Coagulase-Negative Staphylococci," in *Infections Associated with Indwelling Medical Devices*, 3rd ed, ed. F A Waldvogel, A L Bisno (Washington, DC: American Society for Microbiology, 2000), 55–88.
8. J M Anderson, R E Marchant, "Bioma-terials: Factors Favoring Colonization and Infection," in *Infections Associated with Indwelling Medical Devices*, 3rd ed., ed. F A Waldvogel, A L Bisno (Washington, DC: American Society for Microbiology, 2000), 89–109.
9. S E Cramton, F Gotz, "Biofilm Development in *Staphylococcus*," in *Microbial Biofilms*, ed. M Ghannoum, G A O'Toole (Washington, DC: American Society for Microbiology, 2004), 64–84.
10. D Spratt, "Dental Plaque," in *Medical Biofilms: Detection, Prevention and Control*, ed. J Jass, S Surman, J Waller (West Sussex, UK: John Wiley & Sons, 2003), 1–28.
11. P E Kolenbrander, R J Palmer, "Human Oral Bacterial Biofilms," in *Medical Biofilms: Detection, Prevention and Control*, ed. J Jass, S Surman, J Waller (West Sussex, UK: John Wiley & Sons, 2004), 85–117.
12. J G Thomas, G Ramage, J L Lopez-Ribot, "Biofilms and Implant Infections," in *Microbial Biofilms*, ed. M Ghannoum, G A O'Toole (Washington, DC: American Society for Microbiology, 2004), 269–93.
13. J M Steckelburg, D R Osman, "Prosthetic Joint Infections," in *Infections Associated with Indwelling Medical Devices*, 3rd ed., ed. F A Waldvogel, A L Bisno (Washington, DC: American Society for Microbiology, 2000), 173–209.
14. A Stein, M Drancourt, D Raoult, "Ambulatory Management of Infected Orthopedic Implants," in *Infections Associated with Indwelling Medical Devices*, 3rd ed., ed. F A Waldvogel, A L Bisno (Washington, DC: American Society for Microbiology, 2000), 211–30.

15. N Phillips, *Berry & Kohn's Operating Room Technique*, 10th ed. (St. Louis: Mosby, 2000), 740–65.
16. G D Ehrlich, F Z Hu, J C Post, "Role for Biofilms in Infectious Disease," in *Microbial Biofilms*, ed. M Ghannoum, G A O'Toole (Washington, DC: American Society for Microbiology, 2004), 332–58.
17. P S Stewart, P K Mukherjee, M A Ghannoum, "Biofilm Antimicrobial Resistance," in *Microbial Biofilms*, ed. M Ghannoum, G A O'Toole (Washington, DC: American Society for Microbiology, 2004), 250–68.
18. C Walsh, *Antibiotics: Actions, Origins, Resistance* (Washington, DC: American Society for Microbiology, 2003).
19. J Netting, "Sticky Situations," *Science News Online* 160 (July 2001): 2. Also available at http://www.sciencenews.org/articles/20010714/bob12.asp (accessed 24 Jan 2005).
20. P S Stewart, "Multicellular Resistance: Biofilms," *Trends in Microbiology* 9 (May 2001): 204.
21. R Bayston, "Biofilm Infections on Implant Surfaces," in *Biofilms: Recent Advances in Their Study and Control*, 1st ed., ed. L V Evans (Amsterdam: Harwood Academic Publishers, 2000), 117–31.
22. L A Mermel, "Prevention Strategies for Intra-vascular Catheter-Related Infections," in *Infections Associated with Indwelling Medical Devices*, 2nd ed., ed. F A Waldvogel, A L Bisno (Washington, DC: American Society for Microbiology, 2000), 407–25.
23. M R Brunstedt et al., "Bacteria/Blood/Material Interactions. I. Injected and Pre-seeded Slime-Forming *Staphylococcus epidermis* in Flowing Blood with Biomaterials," *Journal of Biomedical Materials Research* 29 (April 1995): 455–66.
24. G W Dentin, "Chlorhexidine," in *Disinfection, Sterilization and Preservation*, 5th ed., ed. S S Block (Philadelphia: Lippincott, Williams & Wilkins, 2001), 321–35.
25. Y Ali et al., "Alcohols," in *Disinfection, Sterilization and Preservation*, 5th ed., ed. S S Block (Philadelphia: Lippincott, Williams & Wilkins, 2001), 229–53.
26. W Goltardi, "Iodine and Iodine Complexes," in *Disinfection, Sterilization and Preservation*, 5th ed., ed. S S Block (Philadelphia: Lippincott, Williams & Wilkins, 2001), 159–84.
27. D W Hobson, L A Seal, "Antimicrobial Bodywashes," in *Handbook of Topical Antimicrobials*, ed. D S Paulson (New York: Marcel Dekker, 2003), 167–88.
28. D S Paulson, "Full Body Showerwash: Efficacy Evaluation of a 4% Chlorhexidine Gluconate," in *Handbook of Topical Antimicrobials*, ed. D S Paulson (New York: Marcel Dekker, 2003), 189–96.
29. D S Paulson, *Topical Antimicrobial Testing and Evaluation* (New York: Marcel Dekker, 1999).
30. S M Nettina, *The Lippincott Manual of Nursing Practice*, 7th ed. (Philadelphia: Lippincott, Williams & Wilkins, 2001), 114–15.
31. D M Fogg, "Infection Prevention and Control," in *Alexander's Care of the Patient in Surgery*, 12th ed., ed. J C Rothrock (St. Louis: Mosby, 2003), 817–930.
32. B Bowen, "Orthopedic Surgery," in *Alexander's Care of the Patient in Surgery*, 12th ed., ed. J C Rothrock (St. Louis: Mosby, 2003), 817–930.

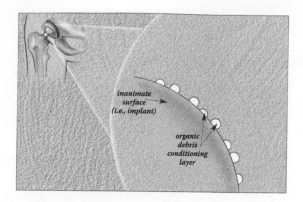

Figure 2.1 Surface conditioning. Organic debris conditions the inanimate surface of implants and other biomaterials. (Reprinted from *AORN*, Vol. 81, No. 3, Daryl S. Paulson, PhD. Efficacy of Preoperative Antimicrobial Skin Preparation Solutions on Biofilm Bacteria, 491–501, 2005. With permission from Elsevier.)

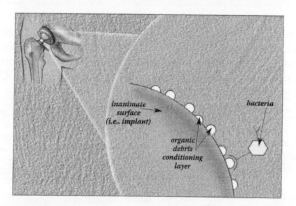

Figure 2.2 Bacterial attachment. Bacteria can attach to the organic debris by chemical adhesion. (Reprinted from *AORN*, Vol. 81, No. 3, Daryl S. Paulson, PhD. Efficacy of Preoperative Antimicrobial Skin Preparation Solutions on Biofilm Bacteria, 491–501, 2005. With permission from Elsevier.)

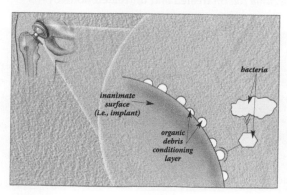

Figure 2.3 Coadhesion. Some bacteria that cannot attach directly to the organic debris can attach to bacteria that can attach to organic debris. (Reprinted from *AORN*, Vol. 81, No. 3, Daryl S. Paulson, PhD. Efficacy of Preoperative Antimicrobial Skin Preparation Solutions on Biofilm Bacteria, 491–501, 2005. With permission from Elsevier.)

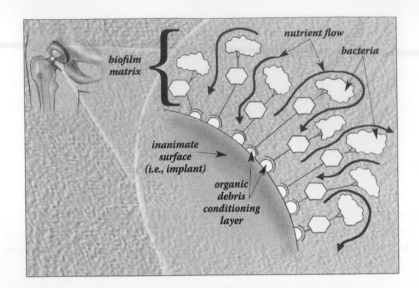

Figure 2.4 Biofilm matrix. The individual bacterial cells form an elaborate matrix of exopolysaccharides and interstitial fluid. (Reprinted from *AORN*, Vol. 81, No. 3, Daryl S. Paulson, PhD. Efficacy of Preoperative Antimicrobial Skin Preparation Solutions on Biofilm Bacteria, 491–501, 2005. With permission from Elsevier.)

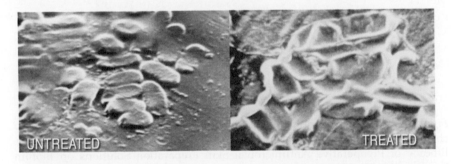

Figure 4.6 *Escherichia coli* on treated and untreated nonwovens.

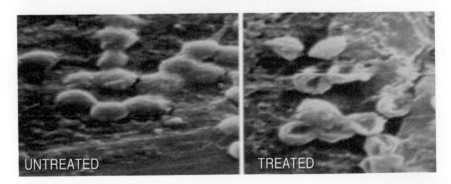

Figure 4.7 *Staphylococcus aureus* on treated and untreated nonwovens.

chapter three

Oral biofilms and transferable antibiotic resistance

Adam P. Roberts

Contents

Introduction: The oral cavity

The oral cavity is one of the most extensively colonized environments in the human body. This is due to the fact that the oral cavity contains many distinct ecological habitats, such as the tooth surface, both above and below the gingival margin; saliva; and the various mucosal surfaces, such as the surface of the tongue. Also, the oral cavity often comes into contact with a wide variety of growth substrates through the ingestion of different foods. The number of species of bacteria contained in this environment is thought to be between five and six hundred [1,2]. Not all of these species are currently able to be cultivated under laboratory conditions. Researchers have estimated that approximately 10 to 50% of these different species of bacteria are culturable [1,3]. In a recent study, approximately 40% of the bacteria identified by amplification and sequencing of the 16S rRNA gene were novel, and probably represent the species that cannot yet be grown under laboratory conditions [1]. More recent estimates, again based on extensive cloning and sequencing of the 16S rRNA genes, have put this figure at approximately one thousand species [4]. In any one oral cavity, however, the typical amount of bacterial species that will be able to be cultured from a single plaque sample is between twenty and thirty. Therefore, the total number of species present in the sample is likely to be between forty and sixty. Additionally, the species profile is likely to be different for different samples investigated [5]. The majority of bacteria in the oral cavity grow on the surface of the tooth as part of a complex, multispecies biofilm commonly known as dental plaque.

Dental plaque

Many, if not most of the bacteria routinely isolated from the oral cavity exist as part of a multispecies biofilm, primarily found on the exposed, nonshedding surface of the tooth. Additional and distinct biofilms also exist, to a lesser extent, on the different mucosal surfaces found within the oral cavity. The biofilms present on the surface of the teeth are commonly known as dental plaque, and they are responsible for two of the most prevalent diseases to affect man, i.e., dental caries and inflammatory periodontal disease [6], which can affect up to 90% of the planet's population [7].

The development of dental plaque begins when the newly exposed surface of the tooth (either following eruption of a new tooth or following mechanical removal of existing plaque by brushing or scaling) is covered, within minutes, by the acquired pellicle. The pellicle is composed of different host-derived molecules, including mucins, proteins, and agglutinins, which act as a source of receptors that are recognized by various oral bacteria. Bacteria that can recognize these receptors and bind to the pellicle-covered surface of the tooth are known as early colonizers and include *Streptococcus* spp., *Actinomyces* spp., *Capnocytophaga* spp., *Eikenella* spp., *Haemophilus* spp., *Prevotella* spp., *Propionibacterium* spp., and *Veillonella* spp. [8]. Coaggregation between many pairs of these early colonizers has been demonstrated experimentally. *Streptococcus* spp. in particular are able to coaggregate with many of the different early colonizers and bind to a variety of different host molecules. This may give them an advantage in establishing themselves on the newly exposed surface of the tooth.

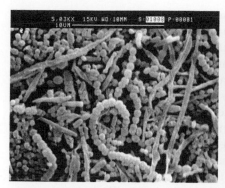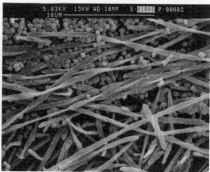

Figure 3.1 Scanning electron micrographs (SEMs) showing the multispecies nature of an oral biofilm. These biofilms were grown in a constant-depth film fermentor for 8 days. Human saliva was used as an inoculum, and the biofilms were fed with artificial saliva. (Courtesy of Dr. Chris Hope, School of Dentistry, University of Liverpool, UK.)

Following colonization by early colonizers, the biofilm is colonized by the late colonizers, including *Actinobacillus* spp., *Prevotella* spp., *Eubacterium* spp., *Porphorymonas* spp., and *Treponema* spp. Coaggregation between pairs of late colonizers is relatively rare compared to those observed for the early colonizers; however, one bacterium acts as a bridge between the two groups. *Fusobacterium nucleatum* is able to coaggregate to all members of both the early and late colonizers examined, and also binds to host-derived molecules found in the pellicle, such as the phosphate-rich protein statherin [8,9].

As the attached bacterial cells grow and divide, many will start to express a biofilm phenotype, which will include the secretion of extracellular polymeric substances (EPS) [10]. These EPS will eventually form the bulk of the mature biofilm structure [11] and afford the cells and microcolonies growing within this matrix physical stability from shear forces experienced within the oral cavity, such as those from the flow of saliva and the action of the tongue and other mucosal surfaces.

Figure 3.1 shows examples of two mature (8-day) oral biofilms grown in a constant-depth film fermentor (CDFF) from an inoculum of human saliva. A variety of different cellular morphologies can be seen in both biofilms illustrating the polymicrobial societies that constitute dental plaque.

Diseases caused by dental biofilms

The organisms present in dental plaque are responsible for a number of different diseases. Acid-producing streptococci such as *Streptococcus mutans* are the primary cause of dental caries. Additionally, plaque contains other pathogens, such as *Tanerella forsythensis*, *Porphyromonas gingivalis*, and *Actinobacillus actinomycetemcomitans*, that have all been shown to be involved in the formation of periodontal disease [12]. Furthermore, oral bacteria can often be isolated from patients with native heart valve endocarditis, as the oral cavity provides an excellent portal to the rest of the body, either as a portal to the alimentary canal or through the gingival surface following cuts or abrasions sustained from brushing or eating, leading to short-term bacteremia. There is also an increasing amount of foreign implants, such as dentures and bridges, being used. These share a common property with the surface of the tooth in that they are non-shedding. Therefore, this surface, once coated with the pellicle, will be readily colonized by bacteria leading to the formation of biofilms on these devices. The presence of biofilms on these devices will obviously compromise the comfort and treatment of the patient, and therefore there is a need to remove or prevent the biofilms from forming.

However, oral biofilms are not all bad. The majority of bacteria living on the surfaces of the mouth do so in symbiosis with the host [7]. The normal oral biofilm at the gingival margin has beneficial effects, as its constant presence provides an exclusion mechanism for incoming and possibly pathogenic microorganisms entering the oral cavity. This is illustrated by the fact that oral disease is usually accompanied by a shift in the populations of the predominant bacteria within plaque when compared to oral health [13].

The use of antibiotics to treat periodontal disease

The treatment of periodontal disease is quite different from the treatment of the majority of other diseases caused by bacterial infections. To date there is not one individual bacterium that can be considered the sole causative agent of the disease. Periodontitis is more complex in that there are a number of different bacteria that are associated with the majority of (but not all) cases of the disease. *A. actinomycetemcomitans* is the bacteria that most closely fulfills Koch's postulates as the etiological agent of periodontitis, and is most often associated with the aggressive juvenile form of the disease [14]. There is also enough data in the literature to allow dentists to consider *P. gingivalis* and *T. forsythensis* as causative agents of periodontitis [12]. However, the disease can occur in the absence of all three of these indicator organisms. Additionally, the presence of other organisms has been reported to be linked to periodontal disease; these include *Prevotella intermedia, Eikenella corrodens, Campylobacter rectus*, and *Capnocytophaga* spp., *Eubacterium* spp., and spirochetes (reviewed in [15] and references therein).

There has been a wide variety of different antibiotics used to treat periodontitis. These have been administered in different doses, in combinations with other antibiotics and used both alone and in combination with mechanical plaque removal [7,14]. Some of the most common antibiotics that have been shown to result in the required levels in gingival crevicular fluid necessary to kill or inhibit bacterial growth are amoxicillin and amoxicillin-clavulanic acid, the tetracyclines, clindamycin, and metronidazole [14]. Dentists (in the UK) prescribe amoxicillin, penicillin, and metronidazole most often, with other antibiotics being prescribed less often, including cephalosporins, tetracyclines, and macrolides [3]; however, due to the polymicrobial nature of oral biofilms, there is likely to be certain species of bacteria present that are resistant to certain antibiotics. Because of this, the outcome of identical treatment regimes may be different for different patients, and it is for this reason that combination therapies, such as metronidazole and amoxicillin, are now becoming increasingly more important [16].

Intrinsic biofilm-mediated antibiotic resistance

Bacteria growing as part of a biofilm often show increased resistance to antimicrobial agents. A recent article [14] gives some examples of the difference between the minimum inhibitory concentrations (MICs) of tetracycline needed to completely abolish growth of a number of organisms growing as either planktonic cultures or in a subgingival biofilm model (Table 3.1).

The reason for this increased resistance of bacteria growing in oral biofilms is multifactorial. The structure of the biofilm itself confers resistance to antibiotic compounds as the EPS surrounding the bacterial cells within the biofilm may exclude the antibiotic from the immediate vicinity of the bacterial cells [17]. A similar effect may occur with compounds that confer resistance, such as β-lactamases, that are

Table 3.1 Differences in the MIC of Various Bacteria Growing in Planktonic Cultures or as a Subgingival Biofilm

Bacteria	Biofilm MIC	Planktonic MIC	Ratio
Actinomyces naeslundii	512–2,048	32–64	16–32
Bacillus coagulans	128–512	128	1–4
Fusobacterium russi	1,024	8	>128
Staphyloccocus intermedius	>1,024	16–256	4 to >64
Streptococcus parasanguinis	>2,048	32	>64
Streptococcus sanguis	512–2,048	16–128	16–32
Veillonella atypical	128–2,048	8–16	16 to >128
Veillonella dispar	>2,048	8–512	4 to >256
Veillonella parvula	≥2,048	8	≥256

Source: Adapted, with the permission of Blackwell Publishing, from [14].

produced and secreted by certain bacteria. If diffusion through the biofilm structure is impeded by the EPS, a high concentration of these compounds will form near the cells producing them, providing a local area of higher resistance. Another property of bacteria growing in biofilms that contributed to their reduced susceptibility to antibiotics is the heterogeneous nature of bacterial growth within the biofilm structure [17]. Because of diffusion being hampered by the EPS, nutrients and chemicals may form gradients within the biofilms. Deep within the biofilm, cells will be less metabolically active than those nearer the surface of the structure. Many antibiotics act against metabolic targets such as protein synthesis (e.g., erythromycin and tetracycline); therefore, if a cell is not metabolically active, it will seem to be transiently resistant to the action of these antibiotics. Additionally, many antimicrobial compounds are actively pumped into the bacterial cells, such as mercury; again, if metabolic activity is low, then a lower amount of the antimicrobial will be pumped across the cell membrane, leading to a reduced susceptibility to that agent. Another feature of the bacterial growth in biofilms that can affect their sensitivity to antibiotics is the biofilm phenotype that many different bacteria exhibit when growing as a biofilm as opposed to planktonically grown cells. There are a multitude of different genes that are either positively or negatively regulated only when the bacteria are growing as a biofilm, and the similar expression of these genes is not seen in planktonically grown cultures. The regulatory networks controlling these genes are likely to control apparatus such as efflux pumps, which will alter the sensitivity to various compounds and antimicrobials [18]. The increasing age of the biofilm also correlates with decreased ability of an antimicrobial agent to kill the cells [19], as it is likely that more EPS will be excreted, further preventing access of a therapeutic agent to the cells within the biofilm [20]. These mechanisms of resistance are summarized in Figure 3.2.

Mechanisms of bacterial-mediated resistance in oral biofilms

Some bacteria are intrinsically resistant to various antibiotics. For example, aerobic bacteria are generally resistant to metronidazole. Other bacteria may lack the target site of the antibiotic, or lack the transport mechanisms needed to import the drug from the extracellular environment. However, treatment of intrinsically resistant bacteria can usually be achieved with an alternative drug. Acquired antibiotic resistance, however, is more problematic. Acquired antibiotic resistance involves the bacterial genome possessing a gene or set of genes that encode specific proteins that will confer resistance to antibiotic compounds. This resistance is common

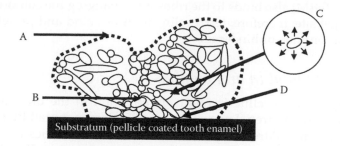

Figure 3.2 Resistance to antimicrobial agents due to the biofilm phenotype. The figure shows cells growing in a multispecies biofilm (unfilled ovoids) surrounded by extracellular polymeric substance (dotted line) adhered to the tooth surface (filled bar). A: The EPS can exclude the antibiotic molecules from penetrating the structure. Additionally, nutrient gradients will also form in this structure due to the EPS. B: Cells can be found within the biofilm structure that are in different metabolic states; therefore, susceptibility to various antimicrobials will differ. C: Any excreted antibiotic resistance compound (such as β-lactamases) (shown as arrows in the blown-up section) may be prevented from diffusing by the EPS, and therefore higher concentrations may occur in some part of the biofilm structure. D: Attached cells may also be exhibiting a biofilm phenotype that can affect susceptibility to antimicrobial agents. Additionally, the age of the biofilm can affect resistance to antimicrobial agents.

and presents an enormous clinical challenge, as the genes encoding these resistance proteins are often present on mobile genetic elements that are capable of transfer from one species of bacteria to another of a different species, or even different genera. Resistance to all of the commonly used antibiotics (in dentistry) has been found in virtually all oral organisms investigated [21].

Resistance to antibiotics falls into one of four different modes of action. Antibiotics may be pumped from cells by the action of efflux pumps. Some of these pumps can act on myriad different compounds; for example, Bmr (*Bacillus subtilis* membrane protein) and Blt (Bmr-like transporter) cause the efflux of a variety of toxins. These include ethidium bromide, rhodamine and acridine dyes, tetraphenylphosphonium, puromycin, chloramphenicol, doxorubicin, and fluoroquinolone antibiotics [22]. Antibiotics can also be altered or inactivated by resistance proteins; this is the case with *tet*(37), a gene isolated from the oral metagenome that confers resistance upon the *E. coli* host (into which it was cloned) by inactivating tetracycline [23]. The target site of the antibiotic can be modified by proteins encoding resistance. This is how Tet(M), a tetracycline resistance protein, encoded by *tet*(M), works [24]. Tetracycline binds to the 30S unit of the ribosome and prevents protein synthesis. Tet(M) also binds to the ribosome, changing the conformation of the target site; therefore, tetracycline does not bind and protein synthesis can continue unhindered.

Resistance found in oral biofilms

Resistance to tetracycline is relatively common in the oral cavity of both adult and children. Lancaster et al. [25] demonstrated that fifteen out of eighteen children (who had not received antibiotics in the preceding 3 months) harbored tetracycline-resistant bacteria. The children were highly unlikely to have come into direct contact with tetracycline, as it is not prescribed for children due to side effects such as the discoloration of the teeth. The most common gene identified in this cohort of children was *tet*(M), which was present on Tn*916*/Tn*1545*-like elements in members of the genera *Streptococcus, Granulicatella, Veillonella,* and *Neisseria* [25]. In an earlier study it was again demonstrated that forty-six out of forty-seven children harbored tetracycline-resistant bacteria; furthermore, it was shown that 11% of the ninety-four different tetracycline-resistant bacteria were resistant to an additional drug. Again, the most common gene found to encode this resistance was *tet*(M), predominantly present within the streptococci [26]. An additional study on the cultivable oral microflora of children has shown that in addition to tetracycline resistance, resistance to ampicillin, erythromycin, and penicillin is also present. All of the antibiotic-resistant bacteria were identified (at least to the genus level), and the resistances found are shown in Table 3.2. This study also demonstrated that resistance to at least two different antibiotics was present in almost 28% of the 432 antibiotic-resistant organisms isolated from the thirty-five subjects [27]. Studies on the oral flora of healthy adults, again who had not received antibiotics for the previous 3 months, also demonstrated the presence of many different tetracycline resistance genes [28], the most common being *tet*(M), followed by *tet*(W), a recently identified gene originally discovered in the rumen anaerobic bacteria *Butyrivibrio fibrisolvens* [29].

Macrolide-resistant bacteria have also been characterized from the oral cavity [27,30,31]. Recently, however, the total cultivable flora was assayed for resistance to erythromycin, and the *mef* and *ermB* genes were shown to be the most common in the samples examined. On average, 7% of the cultivable microflora were resistant to erythromycin and were shown to carry at least one erythromycin resistance gene. Furthermore, the *ermB* gene was shown to be present on Tn*916*/Tn*1545*-like conjugative transposons, which also contain *tet*(M) and *aphA3* conferring tetracycline and kanamycin resistance, respectively [32]. Similar results were found in other studies investigating the molecular basis of resistance

Table 3.2 Number of Isolates Resistant to Antibiotics

Bacterial genus	Amp^R	Erm^R	Pen^R	Tc^R
Actinobacillus spp.	1	3	2	0
Bacillus sp.	0	1	1	0
Bacteroides spp.	1	—	1	0
Capnocytophaga spp.	3	3	7	0
Eubacterium sp.	0	0	0	2
Fusobacterium spp.	0	—	2	0
Granulicatella sp.	0	0	0	1
Haemophilus spp.	37	—	—	7
Leptotrichia sp.	0	—	0	2
Moraxella sp.	0	1	0	0
Neisseria spp.	14	14	10	1
Patsurella sp.	4	—	5	0
Porphyromonas spp.	0	—	2	0
Prevotella spp.	4	—	5	0
Stomatococcus sp.	2	0	0	0
Streptococcus spp.	41	36	37	38
Veillonella spp.	43	—	82	8

Source: Adapted, with the permission of Mary Ann Liebert, Inc., from [27]

Amp^R, ampicillin resistance; Erm^R, erythromycin resistance; Pen^R, penicillin resistance; Tc^R, tetracycline resistance; —, genera not included in antibiotic resistance guidelines.

to macrolides in commensal viridans streptococci and *Gamella* spp. [33–35]. These studies also demonstrated that the majority of strains with an identifiable erythromycin resistance gene contained the *mef* gene. Recently the *mef* gene has been found on a novel conjugative transposon Tn1207.3 in *Streptococcus pyogenes* [36]. Also, the results from these studies show that *ermB* was again linked to Tn*916*/Tn*1545*-like conjugative transposons [35]. The observation that both *mef* and *ermB* are contained within conjugative transposons may go some way to explain their broad distribution. Another study analyzing the effects of treatment with clarithromycin showed that following treatment for approximately 2 weeks, the oropharyngeal flora macrolide resistance had risen from 3.4 to 21.1% of the total cultivable flora, and that this elevated level of resistance was sustained for at least 8 weeks after treatment [37]. This change in the antibiotic resistance profile of oral biofilms following exposure to antibiotics is important and will be considered in greater detail later in this chapter.

As we have seen, many of these resistance genes are found on mobile elements that are capable of transferring from one bacterium to another. This is probably why antibiotic resistance is so common in this environment. The movement of these different mobile elements will be considered in the next section.

Mechanisms of gene transfer among oral bacteria

DNA can transfer from one bacterial cell to another by one or more of three mechanisms: conjugation, transformation, and transduction. Conjugation is the polar transfer of DNA (usually a conjugative plasmid or conjugative transposon) between two bacterial cells, and requires intimate cell-to-cell contact for transfer to occur. Transformation is the process whereby bacteria take up free DNA from the environment and incorporate it into their genome. Bacteria able to take up DNA by this process are termed competent. Finally, transduction is the process in which bacterial DNA gets erroneously packaged into the heads of bacteriophages; when the phage infects another bacterial cell, the packaged DNA is incorporated into the new host's genome. An overview of these mechanisms is shown in Figure 3.3.

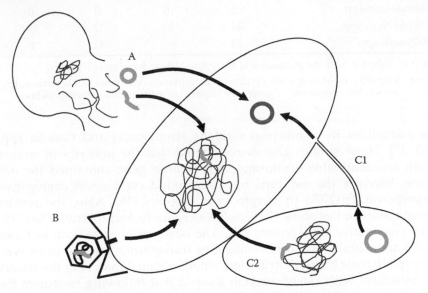

Figure 3.3 Transfer of genetic information between bacterial cells. The mobile DNA is shown as thick gray lines, and chromosomal DNA is shown as thin lines. Cell membranes are thin black lines. A: Transformation—A bacterial cell has lysed and the DNA has been released, where it transforms a competent cell. B: Transduction—a phage particle infects a bacterial cell with its genome, which also contains a region of DNA from the previous bacterial host (thick gray line). C1: Conjugation of a conjugative plasmid from one cell to another; can occur through a pilus. C2: Conjugation of a conjugative transposon, requiring intimate cell-to-cell contact.

Mobile elements involved in the transfer of antibiotic resistance

Plasmids usually exist as independently replicating units; however, on some occasions they will integrate into the bacterial chromosome (Figure 3.3). They are sorted into incompatibility (Inc) groups based on their inability to coexist in the same cell. Plasmids from the same Inc group usually have identical or similar replication/partition systems. Only one plasmid from one Inc group can exist in a cell at a time; if another plasmid enters the cell belonging to the same Inc group, one plasmid will eventually be lost during cell division due to mutual interference of the replication process by the other plasmid, leading to an unequal amount of the two plasmids in the dividing cell. Plasmids are common in both Gram-positive and Gram-negative organisms isolated from the oral cavity.

Conjugative transposons are mobile elements that integrate into their hosts' genome. They encode all of the necessary information for intracellular transposition and intercellular conjugation. Some conjugative transposons have a very broad host range. Tn916, for example, is a paradigm of a large family (Tn916/Tn1545 family) of related promiscuous elements that have been found in or have been introduced into over fifty different species of bacteria (Table 3.3) [38], and it and its close relatives have been found in many oral bacteria, including the *Streptococcus* spp. [39–41], *Veillonella* spp. [42], and *Fusobacterium nucleatum* [43].

Additionally, some oral bacteria, including members of the genera *Neisseria* [44], *Streptococcus* [45], and *Actinobacillus* [46], are naturally competent, and therefore, in theory, any DNA sequence has the potential to

Table 3.3 Different Genera in Which Tn916-Like Elements Have Been Found or Introduced

Acetobacterium	*Haemophilus*
Acholeplasma	*Lactobacillus*
Actinobacillus	*Lactococcus*
Alcaligenes	*Leuconostoc*
Bacillus	*Listeria*
Butyrivibrio	*Mycoplasma*
Citrobacter	*Neisseria*
Clostridium	*Peptostreptococcus*
Enterococcus	*Staphylococcus*
Escherichia	*Streptococcus*
Eubacterium	*Thermus*
Fusobacterium	*Veillonella*

Source: Adapted, with the permission of the American Society for Microbiology, from [38].

be incorporated into the bacterial genome whether it is considered mobile or not. An overview of the movement of the different types of mobile elements and the processes responsible is given in Figure 3.3. This figure shows the movement of DNA by conjugation, mediated by plasmids and transposons, and transduction, mediated by bacteriophages and transformation of DNA derived from dead, lysed bacterial cells.

Evidence for horizontal gene transfer of antibiotic resistance in the oral cavity

Whole genome sequencing and the analysis of individual genes can tell us if particular genes are likely to have been acquired by horizontal transfer, as genes that are 95 to 100% identical in diverse bacteria (such as *tet*(M)) are highly likely to have been acquired in this way. The oral *S. mutans* strain UA159 is naturally competent; i.e., it is able to take up and express both plasmid and chromosomally located antibiotic resistance markers [47]. The completed sequence of the *S. mutans* UA159 genome [48] reveals the presence of a putative conjugative transposon, TnSmu1, that is related to, but distinct from, Tn916. Additionally, there is TnSmu2, an approximately 40 kb region of DNA that contains remnants of insertion sequence elements flanking genes that have the potential to encode gramicidin and bacitracin synthetases. The %G + C value of this region is 28.9%, compared with an overall average of 36.82%, inferring a foreign origin for this DNA. The genome sequence of *Porphyromonas gingivalis* strain W38 has also revealed the presence of a variety of putative mobile elements, including what could be the remnants of conjugative transposons [49]. Additionally, a recent comparative genomics study reports that 35 to 56% (depending on the comparative method employed) of the *F. nucleatum* genome is derived from a foreign origin [50]. It must be remembered, however, that functionality of the majority of putative mobile elements discovered during genome sequencing projects remains to be determined experimentally, and they may in fact be ancient elements that have lost their mobility functions.

Transformation in the oral cavity

Successful transformation of a bacterial cell depends on both physical and chemical factors (e.g., size, conformation, origin and concentration of DNA, UV light, salt, pH, temperature, and nucleases present in the environment) and genetic factors (e.g., the presence of restriction systems in the transformant and the ability of the incoming DNA to either replicate autonomously or integrate into the recipient's genome (reviewed in [51])).

Transformation has no requirement for live donor cells, as the DNA released upon cell death is the principal source of transforming DNA [52].

Therefore, one of the rate-limiting steps for transformation of bacteria growing in an oral biofilm is the longevity of DNA molecules in this environment.

The fate of *Lactococcus lactis* chromosomal DNA, plasmid DNA (pVACMC1 and pIL253), and linear lambda bacteriophage DNA in human saliva has been investigated by incubation with human saliva. The DNA, although partially degraded, was still visible on an agarose gel after 3.5 min. Furthermore, demonstration of the presence of a 520 bp fragment of the pVACMC1 DNA was demonstrated by PCR after 24 h of incubation in human saliva [53]. In another experiment the ability of saliva-incubated pVACMC1 DNA to transform *Streptococcus gordonii* DL1 cells was assayed. The plasmid showed a rapidly decreasing capacity to transform the host to erythromycin resistance with DNA incubated in human saliva for 2 min showing a tenfold decrease in transformation efficiency when compared to untreated DNA [53]. Later experiments looking at the transformation of *S. gordonii* by pVACMC1 incubated in the oral cavity of a human volunteer showed a decrease in transformation efficiency by approximately fourfold when compared to the earlier *in vitro* studies [54]. Another study using plasmid pUC18, incubated in clarified ovine saliva, demonstrated that the plasmid could transform *E. coli* to ampicillin resistance even after 24 h incubation [55]. These studies show that DNA is able to survive in saliva for a long enough period of time to be able to transform suitably competent bacteria present within the oral biofilms.

Additional work in our laboratory has demonstrated that some of the DNA isolated from human saliva is from bacterial organisms other than those usually considered to be members of the oral microflora [56], indicating the presence of DNA from exogenous sources in the oral cavity. This indicates that virtually any DNA that is found free in the oral cavity, whether it is derived from oral bacteria or bacteria present on foodstuffs, is able to be a substrate for transformation. Therefore, it is likely that both interspecies and intergeneric gene transfer can occur by this route within oral biofilms. In a further study, intergeneric plasmid transfer was obtained between *T. denticola* and *S. gordonii* growing both in broth and as a biofilms. These experiments were carried out both with naked plasmid DNA (no donor *T. denticola* cells) and with live *T. denticola* containing the plasmid as a donor. Transformation of the *S. gordonii* cells was observed both in broth and within the biofilms [57].

The effects on competence and transformation of *S. mutans* when growing in a biofilm have been studied. Biofilms, grown on polystyrene plates, of *S. mutans* containing an erythromycin resistance gene were killed by heat treatment. Following this an erythromycin-sensitive strain was allowed to grow as a biofilm on the dead cell-coated plates. All of the six strains of *S. mutans* tested in this study were readily transformable to erythromycin resistance by the DNA of the dead cells [58]. Moreover, transformation frequencies were shown to be ten- to six-hundred-fold higher in biofilm-grown cells than in planktonic cells, indicating that something in the biofilm mode of growth upregulates competence.

Transduction in the oral cavity

One of the main barriers to the activity of bacteriophage in oral biofilms is the access to the cells within the extracellular polymeric substances secreted by the cells themselves when growing as a biofilm [59].

Little is known about the effect of phage and the extent to which transduction contributes to genetic exchange within oral biofilms. Bacteriophage particles have been isolated from human saliva; in one study, three distinct bacteriophages were isolated from enriched human saliva; however, none of the bacteriophages were specific for any of the oral pathogens tested, but did infect *Proteus mirabilis*, also isolated from the saliva. The *P. mirabilis*, and therefore the bacteriophage specific for it, were thought to have been transient in the samples tested, either introduced on food materials or transported manually to the mouth [60]. If cell lysis of the *P. mirabilis* occurred within the oral cavity, obviously this DNA would be available for transforming other naturally competent bacteria. In another study, bacteriophages specific for *E. faecalis* were isolated from the saliva of seven out of thirty-one volunteers [61]. The survival of bacteriophage lambda in saliva has also been determined; bacteriophage lambda dilutions on *E. coli* XL1 revealed that 70% of bacteriophage lambda could be recovered from saliva stored for 1 week at either 4 or 37°C [61].

Additional evidence for the involvement of bacteriophages in the transfer of DNA among residents of oral biofilms comes from studies carried out *in vitro*. For example, tetracycline resistance present on Tn*916* and chloramphenicol resistance present on plasmid pKT210 have been transferred between *A. actinomycetemcomitans* by the generalized transducing phages AaØ23 [62]. Additional evidence that *A. actinomycetemcomitans* bacteriophages may be actively shuttling DNA between bacteria in the oral environment comes from the direct isolation of bacteriophage particles from subgingival plaques from periodontitis patients [62,63].

Conjugation in the oral cavity

There are many examples of conjugative gene transfer reported between species normally found in the oral cavity. Tetracycline-resistant *A. actinomycetemcomitans* has been shown to possess the efflux gene *tet*(B), which was transferable between strains of *A. actinomycetemcomitans* and between *A. actinomycetemcomitans* and *Haemophilus influenzae*. The genetic support for the *tet*(B) gene appears to be a conjugative plasmid [64]. Both chromosomal- and plasmid-borne antibiotic resistance markers have been transferred between oral streptococci in mixed cultures [28,65,66]. Conjugal transfer systems have also been shown to be functional (demonstrated by the transfer of tetracycline resistance) in oral black-pigmented *Bacteroides* spp. [67].

Tn*916*-like elements have been found in many different oral strepto-cocci [68–70], and these have been shown to transfer to other members of this genus during filter-mating experiments [39]. We have shown that con-jugative transposons related to Tn*916* can transfer among the oral strepto-cocci and between the oral streptococci and *E. feacalis* during filter-mating experiments [32,41,42,66]. However, the filter-mating environment does not represent the biofilm environment usually found within the oral cavity.

To address this difference between the conditions found within the oral cavity and the conditions during filter-mating experiments, more com-plex transfer studies have been carried out. The CDFF allows researchers to grow oral biofilms from an inoculum of saliva. It has been used in the past to study different aspects of dental plaque, such as the viability and distribution of cells within plaque [71] and the effects of antibiotics such as tetracycline on plaque [72]. Transient bacteria such as *B. subtilis*, normally considered a soil bacterium, may be able to act as a donor to oral bacteria in the short time that they are present in the oral cavity. We have shown that Tn*5397* (a conjugative transposon carrying *tet*(M)) in a *B. subtilis* donor can transfer to a streptococcal recipient growing as part of an oral biofilm in the CDFF [73]. The *B. subtilis* donor could not be recovered from the biofilms 24 h after inoculation, showing that even though the donor bacte-ria are no longer present, the genetic information contained within them can persist. Further work has demonstrated transfer of native Tn*916*-like elements between oral streptococci grown in the CDFF [41]. However, as DNase was not included in the growth medium fed into the fermentor during these experiments, it is not possible to distinguish between trans-formation and conjugation as the means of DNA transfer.

The effects of antibiotics on horizontal gene transfer within the oral cavity

Antibiotics not only affect the microbial flora in the oral environment, but can also promote the transfer of the elements that encode the resistance to them, as the transfer of some elements, such as Tn*916*, is enhanced by the antibiotic (tetracycline) that they encode resistance to [74]. Ready et al. [72] showed that microcosm dental plaques grown in the CDFF had an altered composition following exposure to tetracycline (Figure 3.4). The levels of tetracycline included in these experiments mimicked that which was found in the gingival fluid following an oral dose of tetracycline [75]. As expected, the biofilm microflora changed to predominantly tetracycline-resistant species (45% of the total). However, following the exposure and during the

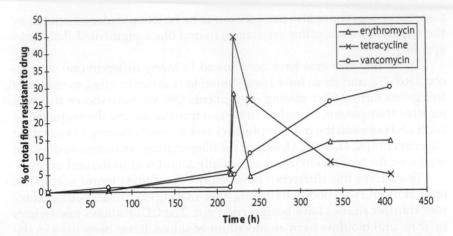

Figure 3.4 Change in antibiotic resistance profiles of oral biofilms following the addition of tetracycline. Tetracycline was added at 214 h. (Adapted, with the permission of Oxford University Press, from [72].)

recovery of the biofilm, it was observed that the biofilms had changed from a community composed predominantly of streptococci to one composed mainly of *Lactobacillus*. This study illustrates that the exposure of oral biofilms to antibiotics does have profound effects on the bacterial composition present within these biofilms. Additionally, the antibiotic resistance profile of the biofilms changed after exposure to tetracycline. There was, for example, a dramatic increase in vancomycin-resistant organisms (1% of the total, increasing to 30% 8 days postdelivery of the tetracycline) due to the intrinsic resistance of the *Lactobacillus*. Additionally, there was a rise in the level of erythromycin resistance during the treatment with tetracycline. This is thought to be due to the linkage of the tetracycline resistance and erythromycin resistance genes on the same element (possibly on Tn916/Tn1545-like conjugative transposons), demonstrating that selection of resistance to one antibiotic can also lead to the selection of resistance genes to other unrelated antibiotics. This linkage of different resistance genes has been recently illustrated by the finding of *Acinetobacter baumannii*, strain AYE, possessing a chromosomal region that contains forty-five different resistance genes to antibiotics, including β-lactams, aminoglycosides, fluoroquinolones, tetracyclines, trimethoprim, chloramphenicol, rifampicin, sulfonamides, and various heavy metals [76].

Concluding remarks

The biofilms found in the oral cavity have been shown to confer resistance to antibiotics by their physical nature and also harbor many different antibiotic-resistant bacteria. Many of the genes conferring this resistance are found on mobile elements that are capable of transferring to new host cells by either transformation, transduction, or conjugation. It is likely that all three mechanisms of gene transfer occur within the oral cavity. However, due to the nature of the usual mode of growth (biofilm), it seems likely that transformation and conjugation contribute more than transduction, although this remains to be determined experimentally. Additionally, it is important to remember that conjugation, transduction, and transformation are not mutually exclusive processes, and situations will occur where a certain fragment of DNA could be transferred by more than one of these three mechanisms. Indeed, the transfer of many mobile elements probably proceeds through bacteria via a combination of these processes.

Acknowledgments

Many thanks to Dr. Chris Hope (University of Liverpool, UK) for images used in Figure 3.1.

References

1. Paster, B.J., et al. Bacterial diversity in human subgingival plaque. *J. Bacteriol.* 183, 3770.
2. Kazor, C.E., et al. Diversity of bacterial populations on the tongue dorsa of patients with halitosis and healthy patients. *J. Clin. Microbiol.* 41, 558, 2003.
3. Sweeney L.C., et al. Antibiotic resistance in general dental practice—A cause for concern? *J. Antimicrob. Chemother.* 53, 567, 2004.
4. Wade, W. Personal communication. 2005
5. Wilson, M. *Microbial inhabitants of humans.* 1st ed. Cambridge, UK: Cambridge University Press, 2005, 333.
6. Wilson, M. Susceptibility of oral bacterial biofilms to antimicrobial agents. *J. Med. Microbiol.* 44, 79, 1996.
7. Pihlstrom, B.L., Michalowicz, B.S., and Johnson, N.W. Periodontal diseases. *Lancet* 366, 1809, 2005.
8. Kolenbrander, P.E., et al. Communication among oral bacteria. *Microbiol. Mol. Biol. Rev.* 66, 486, 2002.
9. Kolenbrander, P.E. Oral microbial communities: Biofilms, interactions, and genetic systems. *Annu. Rev. Microbiol.* 54, 413, 2000.
10. Donlan, R.M. Biofilms: Microbial life on surfaces. *Emerg. Infect. Dis.* 8, 881, 2002.
11. Stephens, C. Microbiology: Breaking down biofilms. *Curr. Biol.* 12, R132, 2002.
12. Socransky, S.S., and Haffajee, A.D. Dental biofilms: Difficult therapeutic targets. *Periodontology 2000* 28, 12, 2002.
13. Marsh, P.D. Plaque as a biofilm: Pharmacological principles of drug delivery and action in the sub- and supragingival environment. *Oral Diseases* 9, 16, 2003.
14. Walker C.B. Kerpina, K., and Baehni, P. Chemotherapeutics: Antibiotic and other antimicrobials. *Periodontology 2000* 36, 146, 2004.

15. Darby, I., and Curtis, M. Microbiology of periodontal disease in children and young adults. *Periodontology 2000* 26, 33, 2001.
16. Slots, J. Selection of antimicrobial agents in periodontal therapy. *J. Periodontal Res.* 37, 389, 2002.
17. Mah, T.-F.C., and O'Toole, G.A. Mechanisms of biofilm resistance to antimicrobial agents. *Trends Microbiol.* 9, 34, 2001.
18. Davies, D. Understanding biofilm resistance to antimicrobial agents. *Nat. Rev. Drug Discov.* 2, 114, 2003.
19. Patel, R. Biofilms and antimicrobial resistance. *Clin. Orthopaedics Related Res.* 437, 41, 2005.
20. Wood, S.R., et al. Changes in the structure and density of oral plaque biofilms with increasing age. *FEMS Microbiol. Ecol.* 39, 239, 2002.
21. Roberts, M.C. Antibiotic toxicity, interactions and resistance development. *Periodontology 2000* 28, 280, 2002.
22. Neyfakh, A.A., Bidnenko, V.E., and Chen, L.B. Efflux mediated multidrug resistance in *Bacillus subtilis*: Similarities and dissimilarities with the mammalian system. *Proc. Natl. Acad. Sci. USA* 88, 4781, 1991.
23. Diaz-Torres, M.L., et al. Novel tetracycline resistance determinant from the oral metagenome. *Antimicrob. Agents Chemother.* 47, 1430, 2003.
24. Speers, B.S., Shoemaker, N.B., and Salyers, A.A. Bacterial resistance to tetracycline: Mechanisms, transfer and clinical significance. *Clin. Microbiol. Rev.* 5, 387, 1992.
25. Lancaster, H., et al. The maintenance in the oral cavity of children of tetracycline resistant bacteria and the genes encoding such resistance. *J. Antimicrob. Chemother.* 56, 524, 2005.
26. Lancaster, H., et al. Prevalence and identification of tetracycline-resistant oral bacteria in children not receiving antibiotic therapy. *FEMS Microbiol. Lett.* 228, 99, 2003.
27. Ready, D., et al. Prevalence, proportions, and identities of antibiotic-resistant bacteria in the oral microflora of healthy children. *Microbiol. Drug Resist.* 9, 367, 2003.
28. Villedieu, A., et al. Prevelance of tetracycline resistance genes in oral bacteria. *Antimicrob. Agents Chemother.* 47, 878, 2003.
29. Scott, K.P., et al. High-frequency transfer of a naturally occurring chromosomal tetracycline resistance element in the ruminal anaerobe *Butyrivibrio fibrisolvens*. *Appl. Environ. Microbiol.* 63, 3405, 1997.
30. Sefton, A.M. Macrolides and changes in the oral flora. *Int. J. Antimicrob. Agents* 11, S23, 1999.
31. Ioannidou, S., et al. Antibiotic resistance rates and macrolide resistance phenotypes of viridans group streptococci from the oropharynx of healthy Greek children. *Int. J. Antimicrob. Agents* 17, 195, 2001.
32. Villedieu, A., et al. Genetic basis of erythromycin resistance in oral bacteria. *Antimicrob. Agents Chemother.* 48, 2298, 2004.
33. Luna, V.A., et al. A variety of gram-positive bacteria carry mobile mef genes. *J. Antimicrob. Chemother.* 44, 19, 1999.
34. Aracil, B., et al. High prevalence of erythromycin-resistant and clindamycin-susceptible (M phenotype) viridans group streptococci from pharyngeal samples: A reservoir of *mef* genes in commensal bacteria. *J. Antimicrob. Chemother.* 48, 592, 2001.

35. Cerda Zolezzi, P., et al. Molecular basis of resistance to macrolides and other antibiotics in commensal viridans group streptococci and *Gemella* spp. and transfer of resistance genes to *Streptococcus pneumoniae*. *Antimicrob. Agents Chemother.* 48, 3462, 2004.

36. Santagati, M., et al. The novel conjugative transposon Tn1207.3 carries the macrolide efflux gene mef(A) in *Streptococcus pyogenes*. *Microb. Drug Resist.* 9, 243, 2003.

37. Berg, H.F., et al. Emergence and persistence of macrolide resistance in oropharyngeal flora and elimination of nasal carriage of *Staphylococcus aureus* after therapy with slow-release clarithromycin: A randomized, double-blind, placebo-controlled study. *Antimicrob. Agents Chemother.* 48, 4183, 2004.

38. Rice, L.B. Tn916 family conjugative transposons and dissemination of anti-microbial resistance determinants. *Antimicrob. Agents Chemother.* 42, 1871, 1998.

39. Hartley, D.L., et al. Disseminated tetracycline resistance in oral streptococci: Implication of a conjugative transposon. *Infect. Immun.* 45, 13, 1984.

40. Bonafede, M.E., Carias, L.L., and Rice, L.B. Enterococcal transposons Tn5384: Evolution of a composite transposon through cointegration of enterococcal and staphylococcal plasmid. *Antimicrob. Agents Chemother.* 41, 1854, 1997.

41. Roberts, A.P., et al. Transfer of Tn916-like elements in microcosm dental plaques. *Antimicrob. Agents Chemother.* 45, 2943, 2001.

42. Ready, D., et al. Potential role of *Veillonella* spp. as a reservoir of transferable tetracycline resistance in the oral cavity. *Antimicrob. Agents Chemother.* 50, 2866, 2006.

43. McKay, T.L., et al. Mobile genetic elements of *Fusobacterium nucleatum*. *Plasmid* 33, 15, 1995.

44. Henriksen, S.D. *Moraxella, Neisseria, Branhamella* and *Acinetobacter*. *Annu. Rev. Microbiol.* 30, 63, 1976.

45. Davidson, J.R., Blevins, W.T., and Feary, T.W. Interspecies transformation of streptomycin resistance in oral streptococci. *Antimicrob. Agents Chemother.* 9, 145, 1976.

46. Fujise, O., et al. Clonal distribution of natural competence in *Actinobacillus actinomycetemcomitans*. *Oral Microbiol. Immunol.* 9, 340, 2004.

47. Murchison, H.H., et al. Transformation of *Streptococcus mutans* with chromosomal and shuttle plasmid (pYA629) DNAs. *Infect. Immun.* 54, 273, 1986.

48. Adjic, D., et al. Genome sequence of *Streptococcus mutans* UA159, a cariodental pathogen. *Proc. Natl. Acad. Sci. USA* 99, 14434, 2002.

49. Nelson, K.E., et al. Complete genome sequence of the oral pathogenic bacterium *Porphyromonas gingivalis* strain W83. *J. Bacteriol.* 185, 5591, 2003.

50. Mira, A., et al. Evolutionary relationships of *Fusobacterium nucleatum* based on phylogentic analysis and comparative genomics. *BMC Evol. Biol.* 4, 50, 2004.

51. Ogunseitan, O.A. Bacterial genetic exchange in nature. *Sci. Prog.* 78, 183, 1995.

52. Cvitkovitch, D.G. Genetic competence and transformation in oral streptococci. *Crit. Rev. Oral Biol. Med.* 12, 217, 2001.

53. Mercer, D.K., et al. Fate of free DNA and transformation of the oral bacterium *Streptococcus gordonii* DL1 by plasmid DNA in human saliva. *Appl. Environ. Microbiol.* 65, 6, 1999.
54. Mercer, D.K., et al. Transformation of an oral bacterium via chromosomal integration of free DNA in the presence of human saliva. *FEMS Microbiol. Lett.* 200, 163, 2001.
55. Duggan, P.S., et al. Survival of free DNA encoding antibiotic resistance from transgenic maize and the transformation activity of DNA in ovine saliva, ovine rumen fluid and silage effluent. *FEMS Microbiol. Lett.* 191, 71, 2000.
56. Seville, L., et al. Unpublished data. 2006.
57. Wang, B.Y., Chi, B., and Kuramitsu H.K. Genetic exchange between *Treponema denticola* and *Streptococcus gordonii* in biofilms. *Oral Microbiol Immun.* 17, 108, 2002.
58. Li, Y.H., et al. Natural genetic transformation of *Streptococcus mutans* growing in biofilms. *J. Bacteriol.* 183, 897, 2004.
59. Sutherland, I.W. The biofilm matrix—An immobilized but dynamic microbial environment. *Trends Microbiol.* 9, 222, 2001.
60. Hitch, G., Pratten, J., and Taylor, P.W. Isolation of bacteriophages form the oral cavity. *Lett. Appl. Microbiol.* 39, 215, 2004.
61. Bachrach, F., et al. Bacteriophage isolation from human saliva. *Lett. Appl. Microbiol.* 36, 50, 2003.
62. Willi, K., et al. Transduction of antibiotic resistance markers among *Actinobacillus actinomycetemcomitans* strains by temperate bacteriophages Aa phi 23. *Cell Mol. Life Sci.* 17, 108, 2002.
63. Sandmeier, H., et al. Temperate bacteriophages are common among *Actinobacillus actinomycetemcomitans* strains isolated from periodontal pockets. *J. Periodontal Res.* 30, 418, 1995.
64. Roe, D.E., et al. Characterisation of tetracycline resistance in *Actinobacillus actinomycetemcomitans*. *Oral Microbiol. Immun.* 10, 227, 1995.
65. Kuramitsu, H.K., and Trappa, V. Genetic exchange between oral streptococci during mixed growth. *J. Gen. Microbiol.* 130, 2497, 1984.
66. Lancaster, H., et al. Characterisation of Tn916S, a Tn916-like element containing the tetracycline resistance determinant *tet*(S). *J. Bacteriol.* 186, 4395, 2004.
67. Guiney, D.G., and Bouic, K. Detection of conjugal transfer systems on oral black pigmented *Bacteroides* spp. *J. Bacteriol.* 172, 495, 1990.
68. Fitzgerald, G.F., and Clewell, D.B. A conjugative transposon (Tn919) in *Streptococcus sanguis*. *Infect. Immun.* 47, 415, 1985.
69. Bentorcha, G., et al. Natural occurrence of structures in oral streptococci and enterococci with DNA homology to Tn916. *Antimicrob. Agents Chemother.* 36, 59, 1992.
70. Lacroix, J.M., and Walker, C.B. Detection and incidence of the tetracycline resistance determinant *tet*(M) in the microflora associated with adult periodontitis. *J. Periodontol.* 66, 102, 1995.
71. Hope, C.K., and Wilson, M. Measuring the thickness of an outer layer of viable bacteria in an oral biofilm by viability mapping. *J. Microbiol. Methods* 54, 403, 2003.

72. Ready, D., et al. Composition and antibiotic resistance profiles of microcosm dental plaques before and after exposure to tetracycline. *J. Antimicrob. Chemother.* 49, 769, 2002.

73. Roberts, A.P., et al. Transfer of a conjugative transposon Tn*5397* in a model oral biofilm. *FEMS Microbiol. Lett.* 177, 63, 1999.

74. Showsh, S.A., and Andrews, R.E. Tetracycline enhances Tn*916*-mediated conjugal transfer. *Plasmid* 28, 213, 1992.

75. Gordon, J.M., et al. Concentration of tetracycline in human gingival fluid after single doses. *J. Clin Periodontol.* 8, 117, 1981.

76. Fournier, P., et al. Comparative genomics of multidrug resistance in *Acinetobacter baumannii. PLoS Genetics*, 2, e7, 2006.

chapter four

Inhibition of foundation colonization of biofilm by surface modification with organofunctional silanes

Robert A. Monticello and W. Curtis White

Contents

Introduction

There are many ways to modify surfaces so that they are less receptive to settling attachment and colonization of microorganisms, or to make such modifications so that microorganisms contacting such surfaces are inhibited or killed or more easily cleaned away. Work with altering surface energies, removing chemical bonding sites by choice of materials or alteration of the surfaces with coatings or similar mollification of the bond sites, providing smooth surfaces to make settling attachment more difficult, producing red-ox surfaces, providing for electrical current transmission or other electrochemical events, and having sacrificial oils or ablative coatings have all been done.

Chemical and physical bonding mechanisms using covalent bonding mechanisms, using covalent or ionic associations done by simple condensation reactions, energy induced as in plasma deposition, or catalyzed reactions of reactive materials can and have been demonstrated. The success of these coatings or surface alterations at controlling the deposition attachment and propagation of good (useful) or bad (destructive, interfering, or annoying) biofilms has often been limited by affective means of attaining durability, being practical for the end use, highly limited by the nature of the target substrate or environmental conditions for successful efficacy, or by cost. Such is the challenge to find technologies that can be evaluated and utilized in a safe, long-lasting, and performance- and cost-effective manner.

The discovery of a group of organofunctional silanes functionalized with cationic moieties by a group of scientists at Dow Corning Corporation, Midland, Michigan, as having the property of both ionically and covalently bonding to receptive surfaces and then homopolymerizing with simple hydrolysis (drying at STP), and that such silanized altered surfaces were antimicrobial in that microbes on contact were killed by the mode of action of rupturing the cells membrane.

Biofilms, for this chapter, are defined as associations of microbes in close proximity to each other within a matrix on surfaces. This allows us to understand these matrices with all of the complexity of their substrates, their binding or separating matrix in layers, swirls, or continuous or discontinuous phases (as they might be emulsions or invert emulsions), nutrient and waste material movements, genetic material exchanges, and with all of the problems and opportunities associated with microbial metabolism.

Silane quaternary ammonium compounds

In the mid-1960s, researchers discovered that antimicrobial organofunctional silanes could be chemically bound to receptive substrates by what were believed to be Si-O linkages. The method was described as orienting the organofunctional silane in such a way that hydrolysable groups on the silicon atom were hydrolyzed to silanols, and the silanols formed chemical bonds with each other and the substrate. The resultant surface modification, when an antimicrobial moiety such as quaternary nitrogen was included, provided for the antimicrobial to be oriented away from the surface [1].

The attachment of this chemical to surfaces appears to involve two processes. First and most important is a very rapid process that coats the substrate with the cationic species one molecule deep. This is an ion exchange process by which the cation of the silane quaternary ammonium compound replaces protons from water on the surface. It has long been known that most surfaces in contact with water generate negative electrical charges at the interface between water and the surface. This mechanism is further supported by data generated with a radioactive silane quaternary ammonium compound. During the treatment, depletion of the radioactivity from solution was almost immediate by an amount corresponding to that sufficient to cover the surface one layer deep, even on surfaces that contain no functionality. Similar results are published for many organic quaternary ammonium compounds. The second process is unique to materials such as silane quaternary ammonium compounds that have silicon functionality, enabling them to polymerize, after they have coated the surface, to become almost irremovable, even on surfaces with which they cannot react. Covalent bonding to that surface will also occur, and it is also possible to have intermolecular polymerization [2] (Figures 4.1 to 4.3).

Works of Abbott, Isquith, Roth, and Walters [3–6] further elaborate the antimicrobial utility of surfaces modified with silane quaternary ammonium compounds. The specific utility of 3-(trimethoxysilyl)

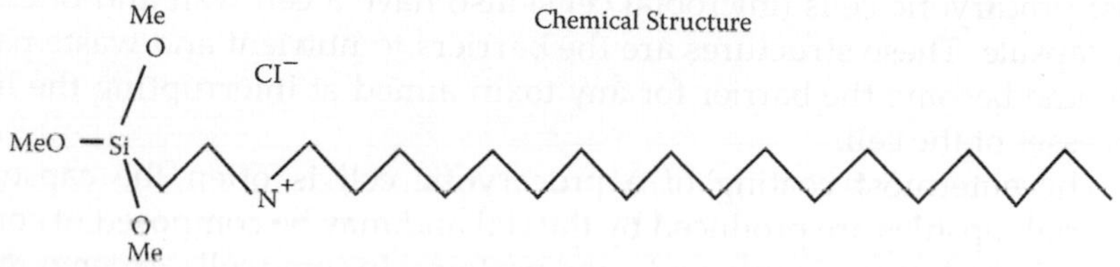

Me
O
/
MeO — Si
 \
  O
  Me

Cl⁻

N⁺

Chemical Structure

Figure 4.1 Organofunctional silane—active ingredient monomer.

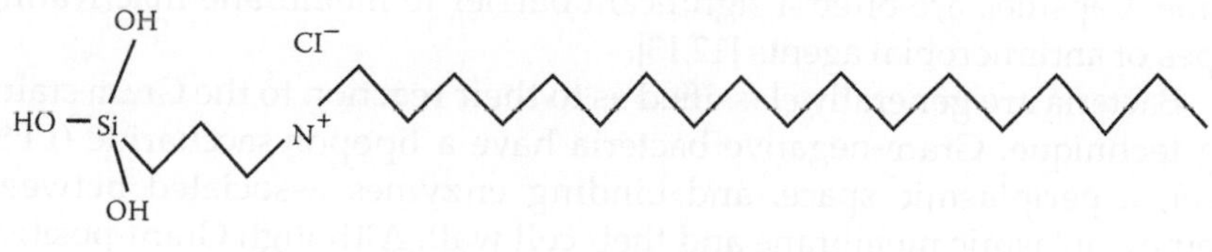

OH
/
HO — Si
 \
  OH

Cl⁻

N⁺

Figure 4.2 Organofunctional silane—hydrolyzed form.

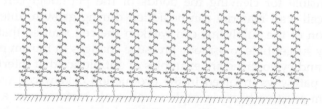

Figure 4.3 Organofunctional silane—polymer network polymerized form.

propyldimethyloctadecyl ammonium chloride (AEGIS™ Antimicrobial, formerly SYLGARD® Treatment based on Dow Corning 5700® Antimicrobial Agent) was described in their work and subsequent works of Gettings and Triplett [7], Speier and Malek [8], Hayes and White [9], McGee et al. [10], and White and Olderman [11].

Microbial surfaces

As indicated above, the clear demonstration of the ability of silane quaternary ammonium compounds to bond or associate with receptive surfaces and to homopolymerize under proper conditions has been demonstrated. The utility of these surfaces in modifying the life processes of microbial cells has also been shown. To better understand the interfaces between the modified surface and the microbial cell, a description of these cell surfaces is appropriate.

Cells are characterized by delineating structures called membranes. Most procaryotic cells (microbial cells) also have a cell wall and often a cell capsule. These structures are the barriers to nutrient and waste passage and become the barrier for any toxin aimed at interrupting the life processes of the cell.

The outermost coating of a procaryotic cell is often the capsule. Bacterial capsules are produced by the cell and may be composed of complex carbohydrates, organic acids, and proteins. In some cells, exoenzymes are associated with capsular materials. Capsules offer a defensive barrier to drying, a storage place for nutrients, and occasionally a site for preparation of nutrients for active or passive movement through the cell membrane. Capsules are often a significant barrier to membrane-inactivating types of antimicrobial agents [12,13].

Bacteria are generally classified as to their reaction to the Gram staining technique. Gram-negative bacteria have a lipopolysaccharide (LPS) layer, a periplasmic space, and binding enzymes associated between their cytoplasmic membrane and their cell wall. Although Gram-positive bacteria do not possess the above features, attached to their cell walls are carbohydrate molecules known as teichoic acids. These molecules are able to regulate the action of enzymes called autolysins. Autolysins are responsible for cleaving the bonds of the peptidoglycan layer so that new cell wall material can be added during growth. Interrupting the function of the autolysins can weaken the cell wall so that the cell is critically susceptible to osmotic pressure changes in the environment. Penicillin and cycloserine are examples of antibiotics that interfere with metabolic pathways involved in the production of new cell wall materials. Gram-positive bacteria replace cell wall materials in a random "patch as needed" process. This and the absence of the teichoic acid–autolysin relationship make them relatively unaffected by antibiotics such as penicillin and cycloserine. These mechanisms, although valuable in bacterial classification, give us little insight into how to design truly broad-spectrum antimicrobials.

The cell walls of procaryotic cells are made of peptidogylcan. Gram-positive bacteria have walls characterized as having a single thick layer of the peptidoglycan. This facilitates space for the lipo-teichoic acid molecules within the peptidoglycan cell wall matrix. Gram-negative bacteria have a thinner, although chemically more complex, wall. This complexity is manifested by the increased branching of the protein chains of the peptidoglycan that links the LPS layer with the plasma membrane.

All components of the cell envelope are important, but the cell membrane has the most complex responsibility for maintaining the life of the cell. The cell membrane must be thought of as a dynamic system. The components are in constant motion, shifting positions while always maintaining the basic sheet-like structure. The components of this system are a phospholipid double layer with hydrophobic ends oriented inwards toward each other. Penetrating and localized to these layers are a great variety of membrane proteins.

Descriptively, one could view this as amorphous masses of proteins moving about on a bed of lipids. Carbohydrates are also associated within and on this dynamic system called the cell membrane. These carbohydrates are dynamically associated with proteins (glycoproteins) and lipids (glycolipids). Adding to the dynamics of the cell membrane are the enzymes, co-enzymes, energy sources such as adenosinetriphosphate, and electrolytes that reside within and on the phospolipid layers. These materials enhance the movement of nutrients and waste materials through processes such as facilitated diffusion, which is a physical phenomenon where nutrients pass through a membrane from areas of high concentration to areas of low concentration by using channels created by facilitating proteins. The simpler unfacilitated physical form of transport is called osmosis and is controlled by the physical nature of the phospholipids of the membrane. Active transport is the movement of nutrients across the membrane with the help of energy-providing molecules such as adenosine triphosphate or nicotinediaminedinucleotide-phosphate, membrane proteins called permeases, and their associated co-enzyme systems.

Association of microorganisms to surfaces

The adhesion of microorganisms to surfaces has been extensively studied since the early works of ZoBell and Anderson [14] and ZoBell [15]. Until the last 15 years or so, efforts were very sporadic. Today a large number of research centers worldwide are contributing heavily to the literature.

Significant to our discussion are several points worth consideration. Bacteria are so small that they behave as colloidal particles [16]. The physiochemical relationship of the bacteria to adhesive and cohesive phenomena helps us partially understand how bacteria associate with our cationic organosilane modified surfaces. The association of macro- and micro-nutrients at the solid-liquid interface often preconditions a surface for microbial colonization. Surface free energy and partitioning capacity are critical phenomena relating to colonization. The dynamics of surface energy change, as nonliving and living fouling occurs, is reviewed thoroughly by Baier [17]. Further, the movement of microorganisms to the nascent or conditioned surface is dependent on active or passive phenomena. Flagellar or ciliary propulsion, as well as chemotactic responses to surfaces, are well understood. Transport of nonmotile (passive) microorganisms to surfaces depends on convection, wave motion, capillary flow, or Brownian movement. Once near a surface, microorganisms are affected by short-range attractive forces such as hydrophobic, coulombic, and van der Waals forces [19]. When sorption is accomplished by these means, viable organisms are readily moved in a so-called reversible sorption [20]. Actual attachment of microorganisms to surfaces is described as adhesion and is established by means of polymer bridging [16]. Permanent adhesion is described as nonspecific when an organism attaches to one site on a surface. It is called specific when interaction between complementary molecular configurations on the surface exists and the organism can move on the surface but separation is resisted. This phenomenon has been described for the gliding bacteria.

The types of forces leading to the attachment of microorganisms to surfaces are influenced strongly by the physicochemical properties of the surfaces of the absorbent and of the microbial cell. Some possible attractive forces would include:

1. Chemical bonding (hydrogen, thio, amide, or ester bonds)
2. Ion-pair formation ($-NH_3^+ \ldots {}^-OOC$)
3. Ion-triplet formation ($-COO^- \ldots Ca^{2+} \ldots {}^-OOC-$)
4. Charge fluctuation
5. Electrostatic attraction between surfaces of dissimilar charge
6. Electrostatic attraction due to image forces
7. Interparticle bridging (polyelectrolytes)
8. Gravitational forces
9. Diffusion forces
10. Charge mosaics
11. Positive chemotaxis (cellular mobility)
12. Surface tension
13. Charge attraction
14. Van der Waals forces of attraction

15. Partitions
16. Hydrodynamic forces
17. Electromagnetic forces

Forces of repulsion would include:

1. Charge repulsion between surfaces of similar charge
2. Van der Waals forces of repulsion
3. Steric exclusion (hindrance)
4. Negative chemotaxis (cellular mobility) [19]

The mechanisms by which microorganisms absorb and adsorb from solid surfaces involve many overlapping phenomena. The attachment or release of cells when they are near solid surfaces may involve a balance between London–van der Waals forces and electrostatic forces of electrical double layers.

The typical microbial cell can be depicted as a macroscopic anion having a large number of electrostatically charged surface sites, or as a nonionigenic surface with hydrophobic areas. An individual cell can participate in specific exchange reactions with synthetic resins, form bonds with water-soluble polyelectrolytes, and attach nonspecifically to other sorbent surfaces. All three of these activities may be manifested during sorption [19].

Antimicrobial activity

The property of the Si-Quat AEM 5772 antimicrobial that provides for the physical contact and rupturing of the cell membranes of single-celled organisms and the association and inactivation of viruses revolves around the chemical structure of the monomer and subsequent final polymer. Contact with the oleophilic moieties of the polymer and the quaternized nitrogen of the polymer by the cell membranes of single-celled organisms causes the physical rupture and inactivation of the membrane and the inhibition and death of the target single-celled organism and the association and inactivation of viruses.

This active ingredient monomer, when treated to surfaces and polymerizes, provides a mode of antimicrobial activity that physically ruptures the cell membranes of microorganisms by ionic association (cell membranes carry a negative charge) and lipophilic attraction (the C18 associating with the lipoprotein of the membrane), causing disruption and lyses of the microbial cell [2] (Figures 4.4 and 4.5). Speier and Malek [8] showed this lysis on treated nonwoven fabric surfaces through electron microscopic observations of model Gram-postive and Gram-negative bacteria (Figures 4.6 and 4.7).

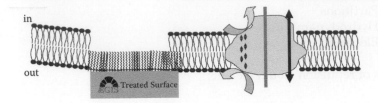

Figure 4.4 Ionic association and lysis.

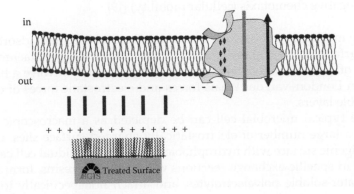

Figure 4.5 Lipophilic association and lysis.

Antimicrobial activity in the prevention of biofilm formation using quaternary amine organofunctional silane technology (Si-Quat)

This section summarizes the broad-spectrum antimicrobial activity of the Si-Quat antimicrobial agent applied onto a variety of both porous and nonporous surfaces. The data represent over 35 years of experience and microbiological and chemical testing measuring the effectiveness of the Si-Quat antimicrobial agent after being applied onto surfaces such as furniture, carpets, wood and vinyl flooring, nonwoven textiles (air filters), aquariums, etc. Surfaces treated with the Si-Quat technology have been shown to be resistant to the formation of biofilm. This resistance is due to two specific mechanisms, which will be described below.

Performance history of silane-based quaternary amine technology

Since inception in the mid-1960s, the antimicrobial activity of the [3-(trimethoxysilyl) propyldimethyloctadecyl] ammonium chloride

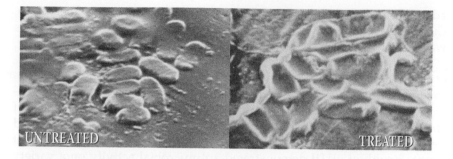

Figure 4.6 (See color insert following page 22.) *Escherichia coli* on treated and untreated nonwovens.

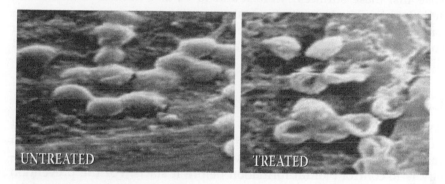

Figure 4.7 (See color insert following page 22.) *Staphylococcus aureus* on treated and untreated nonwovens.

(Si-Quat) has been studied extensively on a variety of treated surfaces. The antimicrobial activity of solid surfaces treated with the Si-Quat agent was first described by Isquith et al. [1] and later elaborated on by others, most notably, by Speier and Malek [2]. In their study, dose-dependent antibacterial activity was demonstrated against both the Gram-negative *Escherichia coli* and the Gram-positive *Staphylococcus aureus* after treating a solid surface of clearly defined dimensions. The rate of kill and surface kinetics of these treated surfaces were further defined and demonstrated by Isquith and McCollum [21]. This work was followed by a companion study that measured the broad-spectrum antimicrobial activity against a mixed fungal spore suspension (*Aspergillus niger, Aspergillus flavus, Aspergillus versicolor, Penicillium funiculosum, Chaetomium globosum*). With the use of radioactive tracers, Isquith and McCollum [21] demonstrated that "biological activity of the Si-Quat bonded to surfaces may offer a method of surface protection without

addition of the chemical to the environment." Algicidal (*Chlorophyta,*
Cyanophyta, and *Chrysophyta*) activity of the Si-Quat applied to glass
was demonstrated by Walters et al. [3].

Further work demonstrates the ability to apply this material to a variety of substrates. This work includes surfaces from glass and aquariums
to entire hospitals [3,22,23]. Kemper et al. [23] studied the microbial colonization of environmental surfaces in hospitals and the effectiveness of
the Si-Quat to control these organisms. This 30-month study measured
persistent antimicrobial activity on surfaces treated with the Si-Quat
agent. Isquith et al. [1] demonstrated antimicrobial activity on a variety
of surfaces. The Si-Quat antimicrobial agent was applied to surfaces as
diverse as stone and ceramic, cotton and wool, vinyl and viscose, aluminum, stainless steel, wood, rubber, plastic, and Formica [1]. These authors
state that these surfaces "were found to exhibit durable antimicrobial
activity when treated with Si-Quat, against a spectrum of microorganisms of medical and economic importance." Further independent testing
confirms antimicrobial activity on air filters and fabrics treated and used
directly in the hospital setting.

Control of biofilm attachment

As outlined earlier in the chapter, biofilm formation is a major cause of
infection, contamination, and product deterioration. Controlling or even
removing the biofilm after its development is difficult. A useful strategy is to control biofilm formation before it starts. For the prevention
of biofilm formation, control of both adherence and colonization of the
microorganisms on the substrate surface is critical. One of the strategies
to prevent biofilm formation is to modify the physiochemical properties of a surface in order to minimize or reduce the attraction of the
surface to the microorganism, thereby controlling adherence. Reducing
the attraction simplistically can be done by either manipulating the ionic
charge of the surface, altering the electrostatic interface, or changing the
hydrophobic/hydrophilic properties through surface energy manipula-
tions (or both) [24].

Controlling or minimizing the adhesion of microorganisms to
the surface can be done using several techniques. Strategies used in
the modification of surface characteristics range from altering the physical properties of the surface via mechanical abrasion to covalently attaching functional components to the surface [25,26]. However, controlling the
physical surface properties through water repellency does not appear to
be enough to prevent biofilm formation. Bacteria can still adhere to highly
hydrophobic surfaces.

Creating an active antimicrobial surface will destroy the adhering
microorganisms, single-celled organisms, thereby preventing further

proliferation. Several groups have recently studied the ability to permanently create antimicrobial surfaces by covalently binding cationic polymers directly to surfaces [27–30].

The idea of creating active antimicrobial surfaces via the treatment with nonleaching quaternary amine compounds is certainly not new, and using very similar approaches to the Si-Quat technology, these groups have created highly active antimicrobial surfaces. Using elaborate application techniques, long polyquaternary chains could be produced that create varied chain length polymers on surfaces with varying thickness. This work is summarized well in the review by Kenawy et al. [28]. These groups demonstrated that a high cationic charge density and specific chain length polymerization were critical in the formation of permanent, nonleaching biocidal surfaces. In theory, these long-chain quaternary polymers are permanently fixed to the surface via covalent linkages but act directly on the cell membrane. This interaction is either through a physical association with the membrane via the long polymeric carbon chains or through direct ion exchange reactions with specific membrane components. The ion exchange theories in particular are interesting with the evidence that high surface charge density is directly related to killing efficiency. The killing efficiency and required charge density are dependent on organism, cellular components, surface charge of particular organisms, and growth rate [31–33].

It is critical, of course, that to use an antimicrobial agent in the prevention of biofilm formation, the agent must be broad spectrum and active against the particular biofilm-causing organisms. Demonstration of the broad-spectrum antimicrobial activity of surfaces treated with the Si-Quat antimicrobial agent can be found in the peer reviewed literature on a monthly basis [8,34]. The Si-Quat technology, as referenced above, is specific against all tested organisms typically responsible for biofilm formation.

Somewhat stimulated by the renewed understanding of the role of Si-Quat-modified surfaces in the prevention of biofilm formation, several investigators renewed the investigation of the relationship between surfaces treated with the ÆGIS Si-Quat chemistry and the formation of microbial biofilm. The application of the ÆGIS Si-Quat onto surfaces structurally changes the surface. To further understand the relationship between water repellency and adsorption on surfaces treated with the Si-Quat, researchers from North Carolina State University, College of Textiles, applied the Si-Quat technology directly onto polyester textiles and measured the water-absorptive properties. This group demonstrated that the siloxane polymer that forms upon final hydrolysis and condensation of the silane monomer is directly related to time, temperature, and pH of treatment solution. Both hydrophilic and hydrophobic surfaces could be created, depending on application procedure [35], while antibacterial activity of the surface remained intact. Saito et al. [36], from Hiroshima University, used treated silica particles to measure the relationship between the adherence of oral streptococci and surface hydrophobicity and zeta potential. Gottenbos et al. [24] from the University of Groningen demonstrated both *in vitro* and *in vivo* activity of Si-Quat-treated silicone rubber used in the biological implants. As an expansion of this work from the same laboratory, Oosterhof et al. [37] measured the inhibitory effects of positively charged coatings on the viability of yeasts and bacteria in mixed biofilm and found a significant reduction in both adherence and colonization of organisms associated with tracheoesophageal shunt prosthetic biofilm.

The Si-Quat technology when applied to surfaces both affects the adhesion properties of microorganisms, due to increased hydrophobic properties of the long carbon chain fully polymerized, and directly destroys one-celled organisms on contact through mechanisms described above. Nikawa et al. [38] from Hiroshima University studied both the adhesion and colonization of mixed biofilm suspensions as a means to control biofilm formation on medical devices. This group demonstrated that commercially pure wrought titanium treated with the Si-Quat technology significantly reduced the adherence and colonization of both *Candida albicans* and *Streptococcus mutans*, even when the surface was coated with a proteinaceous layer like saliva or serum. Clearly this biofilm control mechanism was directly related to the decreased adhesion due to the hydrophobicity created by the octadecyl alkyl chain, and also due to the killing of the quaternary ammonium, which killed initial adherent cells and also retarded or inhibited subsequent microbial growth. Furthermore, cell culture and cytotoxicity studies were performed in order to demonstrated the nonleaching behavior of the antimicrobial coating. No significant cytotoxicity of Si-Quat was observed in cell viability tests or inflammatory assays.

This chapter is really just at the beginning. This area of investigation and its relationship to biofilm formation and control has only just begun, and there are a number of active research centers looking at this set of technologies and its effects on a wide variety of substrates and biofilm combinations.

References

1. Isquith, A.J., et al, *Appl. Microbiol.* (1972): 24(6): 859–63.
2. Malek, J.R., and Speier, J.L., Development of an organosilicone antimicrobial agent for the treatment of surfaces, *J. Coasted Fabrics* 12 (1982): 38–46.
3. Walters, P.A., et al., *Appl. Microbiol.* 25 (1973): 256.
4. Roth, C., Canadian Patent 2,010,782 (May 24, 1977).
5. Abbott, E.A., et al., U.S. Patent 3,817,739 (June 18, 1974) and U.S. Patent 3,865,728 (February 11, 1975).

6. Abbott, E.A., and Isquith, A.J., U.S. Patent 3,794,736 (February 26, 1974).
7. Gettings, R.L., and Triplett, B.L., A new durable antimicrobial finish for textiles, Book of Papers, AATCC National Conference (1978).
8. Speier, J.L., and Malek, J.R., Destruction of microorganisms by contact with solid surfaces, *J. Colloid Interface Sci.* 89 (1982): 68–76.
9. Hayes, S.F., and White, W.C., How antimicrobial treatment can improve nonwovens, *American Dyestuff Reporter* (June 1984).
10. McGee, J.B., et al., New antimicrobial treatment for carpet applications, *American Dyestuff Reporter* (June 1983).
11. White, W.C. and Olderman, G.M., Antimicrobial techniques for medical nonwovens: A case study, Book of Papers, 12th Annual Technology Symposium (May 22–23, 1984).
12. Lawrence, C.A., Germicidal properties of cationic surfactants, in *Cationic surfactants* (New York: Marcel Dekker, 1970), chap. 14.
13. M.E. Ginn, Adsorption of cationic surfactants on miscellaneous solid substrates, in *Cationic surfactants* (New York: Marcel Dekker, 1970), chap. 11.
14. Zobell, C.E., and Anderson, *J. Bacteriol.* 29 (1935): 239.
15. Zobell, C.E., *J. Bacteriol.* 46 (1943): 39.
16. Marshall, K.C., *Interfaces in microbial ecology* (Cambridge, MA: Harvard University Press, 1976), 5, 44.
17. Baier, R.E., Substrate influences on adhesion of microorganisms and their resultant new surface properties, in *Adsorption of microorganisms to surfaces* (New York: John Wiley & Sons, 1980).
18. Young, L.Y., and Mitchell, R., in *Proceedings of Third International Congress of Marine Corrosion and Fouling* (Evanston, IL: Northwestern University Press, 1973), 617.
19. Daniels, S.L., Mechanisms involved in sorption of microorganisms to solid surfaces, in *Adsorption of microorganisms to surfaces* (New York: John Wiley & Sons, 1980).
20. Marshall, V. et al. *J. Gen. Microbiol.* 68 (1971):337.
21. Isquith, A.J. and McCollum, J.C. Surface kinetic test method for determining rate of kill by an antimicrobial solid. *Appl. Environ. Microbiol.* (1978) 36(5): 700–4.
22. Lewbart, R. et al., Safety and efficacy of the environmental products group masterflow aquarium management system with AEGIS Microbe Shield, *Aquacultural Eng.* 19 (1999): 93–98.
23. Kemper, R.A. et al., Improved control of microbial exposure hazards in hospitals: A 30-month field study, paper presented at the National Convention for Association of Practitioner for Infection Control (APIC) 1992.
24. Gottenbos, B. et al., Antimicrobial effects of positively charged surfaces on adhering Gram-positive and Gram-negative bacteria, *J. Antimicrob. Chemother.* 48 (2001), 7–13.
25. Gottenbos, B. et al., In vitro and in vivo antimicrobial activity of covalently coupled quaternary ammonium silane coatings on silicone rubber, *Biomaterials* 23 (2002): 1417–23.
26. MacKintosh, P. et al., *Effects of biomaterial surface chemistry on the adhesion and biofilm formation of staphylococcus epidermidis in vitro* (New York: Wiley Inter-Science, 2006), 836–42.
27. Bouloussa, O. et al., A new, simple approach to confer permanent antimicrobial properties to hydroxylated surfaces by surface functionalization, *Chem. Commun.* (2008): 951–53.

28. Kenawy, E.R. et al., The chemistry and applications of antimicrobial polymers: A state-of-the-art review, *Biol. Macromolecules* 8 (2007): 1359–84.
29. Huang, J. et al., Nonleaching antibacterial glass surfaces via "grafting onto": The effect of the number of quaternary ammonium groups on biocidal activity, *Langmuir* 24 (2008): 6785–95.
30. Lin, J. et al., Insights into bactericidal action of surface-attached poly (vinyl-N-hexylpyridinium) chains, *Biotechnol. Lett.* 24 (2002): 801–5.
31. Kurt, P. et al., Highly effective contact antimicrobial surfaces via polymer surface modifiers, *Langmuir* 23 (2007): 4719–23.
    Murata, L. et al., Permanent, non-leaching antibacterial surfaces. 2. How high density cationic surfaces kill bacterial cells, *Biomaterials* 28 (2007): 4870–79.
32. Kugler, R. et al., Evidence of a charge-density threshold for optimum efficiency of biocidal cationic surfaces, *Microbiology* 151 (2005):1341–48.
33. Neu, T.R., Significance of bacterial surface-active compounds in interaction of bacteria with interfaces, *Microbiological Reviews* (March 1996), 151–66.
34. Andresen, M. et al., Nonleaching antimicrobial films prepared from surface-modified microfibrillated cellulose, *Biomacromolecules* 8 (2007): 2149–155.
35. Abo El Ola, S. et al., Unusual polymerization of 3-(trimethoxysilyl)-propyldimethyloctadecyl ammonium chloride on PET substrates, *Polymer* 45 (2004): 3215–25.
36. Saito, T. et al., Adherence of oral streptococci to an immobilized antimicrobial agent, *Arch. Oral Biol.* 42 (1997): 539–45.
37. Oosterhof, J. et al., Effects of quaternary ammonium silane coatings on mixed fungal and bacterial biofilms on tracheoesophageal shunt prostheses, *Appl. Environ. Microbiol.* (2006): 3673–77.
38. Nikawa, H. et al., Immobilization of octadecyl ammonium chloride on the surface of titanium and its effect on microbial colonization in vitro, *Dental Mater. J.* 24 (2005): 570–82.

chapter five

Biofilms and device implants

Elinor deLancey Pulcini and Garth James

Contents

Introduction

Microorganisms attach to surfaces and form biofilms (Figure 5.1). From an historical perspective, biofilms first became an accepted entity as well as an accepted problem in environmental settings, particularly in industry. The Center for Biofilm Engineering at Montana State University–Bozeman in 1990, funded through the National Science Foundation's

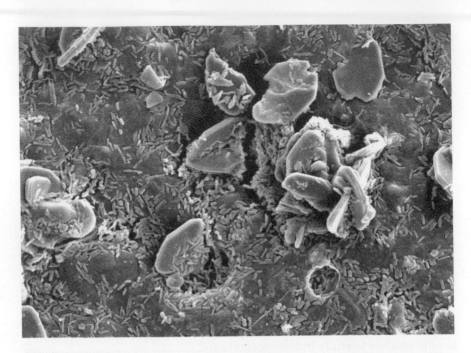

Figure 5.1 Scanning electron micrograph (SEM) of *E. coil* growing in a flow through urinary catheter model system. Image shows clusters of rod-shaped bacterial cells. Larger objects are crystals deposited on the tubing surface.

Table 5.1 Summary of Scirus Search for Articles Containing the Search Term *Biofilm* Compared with the Search Term *Medical AND Biofilm*

Years	Biofilm	Medical AND biofilm	Percent of medical AND biofilm articles compared to biofilm
1980–1985	119	8	7%
1985–1990	411	51	14%
1990–1995	1,086	176	19%
1995–2000	3,754	581	18%
2000–2005	8,376	1,579	23%
2005–2006	4,354	1,027	31%

Engineering Research Center Program, was established primarily to study environmental biofilms. The fact that biofilms were growing and clogging water pipes and cooling towers was easily accepted because the effects of the biofilms were readily apparent in the reduction and eventual loss of flow through a system.

The existence of biofilms in the medical setting has taken somewhat longer for acceptance. Using the Scirus database (http://www.scirus.com/srsapp/) to compare the search term *biofilm* with the search term *medical*

AND biofilm (abstracts and articles in any format) shows that while articles using the word *biofilm* increased dramatically from 1980 to 2006, articles for the use of the term *medical AND biofilm* has lagged (Table 5.1). During the 5-year periods from 1990 to 1995 and from 1995 to 2000, approximately 18 to 19% of the biofilm articles published contained the terms *medical* and *biofilm*. The articles containing the terms *medical* and *biofilm* published from 2000 to 2005 rose slightly to 23%. What is most exciting, however, is that the percentage of articles with the terms *medical* and *biofilm* published in 2005–2006 has increased to 31% of all biofilm articles published for that 1 year. While this is not a formal statistical analysis of publications, it does show a trend toward an increasing awareness and acceptance of the presence of medical biofilms. This increased rate of medical biofilm articles indicates two phenomena. First is an increased acceptance of the role of biofilms on medical devices and in chronic infections by the medical community. Second is the actual increase in rate of biofilm-associated device infections due to the increased use of the devices.

With advances in medicine comes a corresponding increase in the number of medical devices that are placed in patients. These can be short-duration devices, such as endoscopes and arthroscopes or urinary catheters placed for surgery; long-term devices, such as central venous catheters placed for chemotherapy or total parenteral nutrition; or permanent devices, such as artificial hips.

The baby boomer generation, which has already impacted school size and job availability, is now starting to impact the medical establishment with problems associated with aging. As this population ages, it is experiencing an increase in immune suppression rates, which is a normal effect of increasing chronological age. Couple that with increased needs for medical devices (short term to long term) and the environment is perfect for an increase in biofilm-related infections.

In this chapter, we will focus on *bacterial* biofilms on medical devices. This is not to imply that fungal biofilms or fungi in mixed-species biofilms are not medically relevant, but only that we have not studied fungal biofilm formation as extensively as bacterial biofilms.

The statistics on device-related infections

The Centers for Disease Control and Prevention estimated for 2002 that hospital-acquired infection accounts for 2 million infections, 90,000 deaths, and $4.5 billion in excess health care costs annually [1–3]. In a study undertaken at Cook County Hospital, Illinois, published in 2003, the estimated average cost of health care for patients with a suspected hospital-acquired infection increased by $6,767. For those patients with a confirmed hospital-acquired infection, the cost increased to $15,275 [4].

The reporting of hospital-acquired infection rates may not always delineate the types of infections. However, hospitals are starting to change their reporting policies. If the rates for the entire hospital are assessed rather than just the ICU, there are an estimated 250,000 cases annually of bloodstream infections due to infected central venous catheters, with costs estimated in the millions to billions of dollars for treatment [5]. Pennsylvania hospitals reported that of the total 11,668 hospital-acquired infections in 2004, 6,139 were urinary tract infections, 1,317 were surgical site infections, and 945 were multiple infections [6].

Certainly, not all these infections are biofilm infections on devices. However, such device infections may take longer to develop than an average hospital stay. One estimate is that there are 1.32 million prosthetic devices that become infected in the United States each year [7]. For example, of the more than 3 million permanent pacemakers and approximate 180,000 implanted cardiac defibrillators worldwide, it is estimated that the infection rates on these devices range from 0.13% to almost 20% [8].

Bacterial growth characteristics in a biofilm

Bacterial attachment to a surface is a complex, multistep process. Often, bacterial cells will establish a transient attachment to a surface prior to becoming irreversibly attached. As cells attach to the surface, they begin to divide and form microcolonies, which will eventually lead to the development of a mature biofilm [9].

Surface-associated changes are rapid. Proteomic analysis of *Pseudomonas aeruginosa* revealed that protein expression can change in as little as 10 min after initial attachment to a surface. In the span of 3 h post-attachment, the expression of approximately twenty proteins is changed [10]. Most of the proteins fall into the category of DNA regulation and carbon metabolism, which indicates the cell is preparing itself for the lifestyle change from planktonic to sessile.

Biofilm formation is a complex process in which there appears to be no single biofilm gene or operon. A number of studies have searched for the holy grail of biofilm formation, the biofilm phenotype, the existence of which would have great value as a diagnostic tool. Thus far, the results of proteomic and genomic analyses of biofilm vary according to species and biofilm formation conditions [11,12].

These studies illustrate the inherent adaptability of microorganisms. A study of sixty-seven clinical isolates of *Staphylococcus epidermidis* recovered from medical devices showed a range of genetic expression with respect to biofilm formation capability. Some of the clinical isolates capable of biofilm formation possessed the genes previously implicated in biofilm formation in this species, the *ica*ADBC operon. However, other isolates, also capable of biofilm formation, lacked the operon, and some of

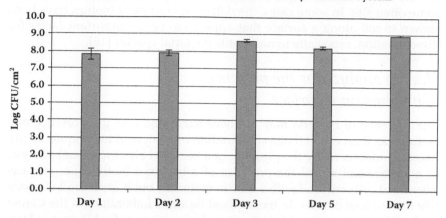

Figure 5.2 Growth of attached *E. coli* in a flow through urinary catheter model system. Prior to sampling, the catheter sections were rinsed to remove planktonic and loosely attached cells. Biofilms were disrupted by vortex, sonication and vortex, and the cell suspension was placed for viable cell counts.

the isolates shown to produce low quantities of polysaccharide intercellular adhesion (PIA) contained the operon [13].

Under laboratory conditions and depending on the bacterial species, incubation conditions, and medium used, bacteria can attach to a surface and form a biofilm with viable cell counts of 10^4 to 10^8 cfu/cm^2 in 24 h. Over time, biofilms achieve a quasi-steady population state in which growth and attachment rates are equal to the rate of detachment. Growth curves of *E. coli* grown in artificial urine over a 7-day period show that the biofilms achieved a steady population of 10^8 cfu/cm^2 (Figure 5.2) [14]. With no change in parameters such a flow rates and nutrient and oxygen availability, the biofilm can likely maintain this population indefinitely. Changes in the environment can result in detachment or sloughing of a substantial portion of the biofilm, or can result in the dispersal of single cells or shedding from the matrix surface.

Detachment

Detachment of a portion of the biofilm can have serious ramifications for the transfer of infection throughout the body. The detached biofilm cell clusters retain the characteristics of a biofilm, including antimicrobial resistance and cohesiveness, because the bacterial cells are still encased in a matrix material. Using time-lapse microscopy of an *in vitro* catheter model of *S. aureus* biofilms, the extent of antibiotic resistance was directly proportional to the size of the detached biofilm particle or embolus [15].

In addition, these cell clusters are, in a sense, primed for reattachment at some other site. In some cases, the detached clumps of biofilm have been shown to roll along a surface that appears to be an important factor for dissemination, especially in nonmotile bacterial species [16].

How to analyze for the presence of biofilms on a surface

The presence of bacteria may not necessarily indicate the presence of a biofilm. So, it is crucial that a combination of analytical techniques be utilized in order to verify the presence of a bacterial biofilm. Visual analysis of a surface can reveal the presence of a bacterial biofilm (i.e., surface-associated cells covered with a matrix material), but it is difficult to quantify the number of cells. In the Medical Biofilms Laboratory at the Center for Biofilm Engineering, Montana State University, we find that use of traditional microbiological quantification techniques combined with visual analysis of the surface is the best means to both quantify the cells in a biofilm and show that a biofilm is indeed present on the surface.

For viable cell counts (plate counts) to be a reasonable estimation of the number of cells present on the surface, the biofilm must be removed and then disrupted without killing the cells in the process. Methods for removal include vortexing, sonication, and scraping, or combinations of these methods. (*Note*: Extreme care must be taken to avoid the production of aerosols when vortexing and sonicating, especially when processing clinical samples.) Cells subsequently suspended in sterile solution are then serially diluted and plated onto an appropriate medium and incubated. Final counts of the cells are expressed as cell forming units per square centimeter (cfu/cm^2) of surface area sampled. Controls can be used to estimate biofilm removal and disaggregation efficiency.

Visualization of the biofilm is done by either epifluorescent microscopy, confocal scanning laser microscopy (CSLM), or scanning electron microscopy (SEM). All three visualization methods have advantages and disadvantages. Both microscopy and CSLM combined with stains or antibody-specific probes can effectively visualize cells within a biofilm. Physiologic stains designed to indicate cell viability can indicate effects of treatments (especially when used in conjunction with viable cell counts). Fluorescent *in situ* hybridization (FISH) probes, which bind specific DNA sequences, can be used to fluorescently label cells of specific species of bacteria within a biofilm. Epifluorescent microscopy is best used on thinner biofilms; however, if the biofilm is patchy, it can become problematic to search for and find a sparse biofilm on a surface. CSLM is best used on thicker biofilms in order to utilize the sectioning capability of this microscope. Another advantage to CSLM imaging is the amenability to the use of computer image processing tools. CSLM can take a series of images through the layers of the biofilm at predetermined thicknesses. This series can be recombined into a three-dimensional image, which can give important information on biofilm thickness and heterogeneity, as well as cell position within the biofilm matrix. With SEM, it is possible to visualize small cells and patches of biofilm on the surface of a device that may not be easily detected with microscopy or CSLM. In addition, SEM can visualize thin biofilms on a surface, whereas CSLM relies more on the presence of thicker biofilms in order to be an effective tool. SEM resolution capabilities allow for the visualization of single cells on a surface. A disadvantage to SEM lies in the processing, which involves vacuum desiccation and sputter coating of the surface with silver or gold. This flattens the biofilm, and thus reduces the information available regarding the spatial heterogeneity of the biofilm, and may introduce artifacts. SEM also only allows only for visualization of the outer surface of the biofilm. Variable-pressure SEM enables the visualization of samples under low-vacuum conditions as well as elemental analysis of the surface. This has proven to be a useful analytical tool, for example, in the examination of the crystals on a urinary catheter. Environmental SEM reduces the amount of processing required, thus allowing for imaging of fully hydrated samples at high resolution.

Surfaces

As yet, there is no known surface to which bacteria cannot attach. Biofilm literature regarding surfaces examined includes a variety of materials and polymers, including metals (stainless steel, iron, copper), polymers (silicone, polystyrene, polycarbonate), glass (glass wool, slides), cement, paint on paintings, and minerals (goethite, hydroxy apatite) [17–25]. Some bacterial species do appear to preferentially attach to surfaces; this can vary by bacterial species tested and the type of surface tested. For example, initial adhesion and biofilm formation on various surfaces by *S. epidermidis* was found to vary by strain with no consistent phenotype [26].

Bacterial detachment or strength of attachment may vary according to surface roughness. *P. aeruginosa*, for example, will attach to either glass slides or hydroxy apatite (HA)–coated slides, growing biofilms that are similar in both cases in terms of length of time to form, thickness, and cfu/cm^2. However, biofilms formed on glass slides are easily removed, whereas those formed on HA slides will still have bacteria present on the surface even after scraping of the slide [27]. Rough surfaces have been shown to adsorb more proteins (i.e., conditioning layer) as well as more bacteria [28]. While surface characteristics such as hydrophobicity may play a role in initial adhesion of bacteria to that surface, these surface characteristics may be rather quickly ameliorated by chemical conditioning layers formed on the surface by proteins or host immune cells in the human body. Once the conditioning layer is established, much of the surface characteristics have been substantially altered, thus changing the surface environment "seen" by the bacterial cells. A survey of nine clinical strains of *S. epidermidis* showed no relationship between the cell surface hydrophobicity and the ability of the cell to preferentially bind to hydrophobic or hydrophilic surfaces [26]. Surface characteristics appear to determine the chemical composition of the conditioning layer. For example, hydrophobic surfaces adsorb more albumin [29]. More agglutinins and proline-rich proteins attached to rough surfaces than to smooth surfaces in a study to assess the effects of surface characteristics on *in vitro* biofilm formation by oral bacteria [28]. These subtle differences in conditioning film composition may affect the types of bacteria that ultimately attach to that surface as well as the patterns of attachment [30].

Devices implanted in the body are coated with a layer of fibrin and fibronectin in a relatively short period of time. Urinary catheters can become coated with proteins and electrolytes to the point that the catheter surface may contain crystalline encrustations along with the biofilm [30].

Cohesion

Cohesion or "stickiness" of the biofilm matrix is a physical aspect of biofilm formation and maintenance that needs to be further studied. Exopolysaccharide (EPS) polymers produced by biofilm bacteria can vary in composition according to carbon availability, bacterial species, and environmental cues such as stress [32]. The stickier the biofilm matrix, the more firmly attached are the cells within the biofilm. The extent of cohesive characteristics within a biofilm matrix has potential implications for the ability of detached clumps of biofilm to efficiently withstand challenge by antibiotics or host immune cells, and thus be able to spread infective bacterial cells from a biofilm-impacted device to other sites in the body [33].

Bacterial strains

Care must be taken regarding the use of bacterial strains to assess biofilm formation and product efficacy in the laboratory setting. Strains that have been subcultured numerous times tend to lose virulence [34]. For medical biofilm work, it is important that stains of bacteria obtained from a clinical setting and minimally subcultured be used in order to fully characterize biofilm formation by clinically relevant bacterial species that still retain much of their virulence expression. A study of the reduction of bacterial genome size using *Salmonella typhimurium* passaged at least twenty times indicated that large-scale genome reduction can take place in a relatively short period of time. Results showed that the bacteria were discarding DNA that was no longer functional for their environment. In this serial passage experiment, deletions in DNA ranged in size from approximately 1,200 base pairs (bp) to over 170,000 bp [35]. A study of diversity over time in *P. aeruginosa* clones grown in drip flow reactors showed phenotypic diversity in 2 to 7 days of biofilm growth. Phenotypic variation was found to increase with increased biofilm growth duration [36]. This phenomenon was observed in *Pseudomonas aeruginosa* cultured from cystic fibrosis patient lungs, which show an emergence of mucoid variants as well as small-colony variants [37].

In vitro *models to study biofilm formation*

There is no model system that is an exact replicate of the system being modeled. Despite this limitation, it is still possible to develop a model that can provide valuable data in a short period of time. The use of *in vitro* models is the crucial first step in the process of the evaluation of a product, which then progress to animal models and finally to clinical trials.

A number of different model types have been developed in order to grow reproducible biofilms. A good *in vitro* model must show minimal variation of biofilm formation criteria, such as bacterial cell counts within a run, and also from one experimental run to another. It is also important that the model be easily manipulated by the technician for treatments and analysis. The more complex the model, the more chance there is for the introduction of contaminating bacteria. It is also important that the surfaces on which the biofilm is grown be easily removed and used for microscopic analysis.

Different model types using the same bacterial strains, media, growth conditions such as temperature, and biofilm formation duration will grow different types of biofilm that respond differently to analysis and treatment.

With any model it is important to have matched controls. Despite the caveat for the need of minimal run-to-run variation, there can still be enough variation that matched controls must be utilized within each experimental run. It is best to use an untreated control, which provides information as to how well the biofilm formed in that particular run, as well as treatment controls. If, for example, the treatments being tested are dissolved in a specific carrier or solvent, then a treatment control of that carrier or solvent without any additional treatment compounds must also be included.

In vitro biofilm models based on the ninety-six-well format are able to generate data in a relatively short period. This format allows for numerous replicate samples utilizing small volumes, so these are excellent choices for the preliminary screening of compounds in which the compounds are expensive or difficult to make. A disadvantage to this model is that it is such a small-volume system that it may be affected more by variations in such parameters as temperature, oxygen, or shaking.

The basic ninety-six-well format has been used successfully [38,39]; however, a criticism of the model is the fact that the biofilm is formed on the bottom on the wells. This means the biofilm can contain both biofilm bacteria and bacterial cells that settled on the bottom of the well as a result of sedimentation.

The minimal biofilm eradication concentration (MBEC) model was designed to overcome some of these limitations [40]. It is based on the ninety-six-well system, in which the top plate contains ninety-six pegs and the bottom plate can be either a trough or a ninety-six-well plate. Advantages to this system are that biofilm tested is attached to the pegs, so it is relatively ensured that the tests are performed on attached cells. In addition, the lid can be easily placed into a fresh plate for media changes.

Disc reactors have proven to be a useful model for growing biofilms under flowing conditions. Two model types are commonly used, the rotating disc reactor (RDR) and the CDC reactor [41,42]. The RDR reactor has spaces for six removable coupons, while the CDC reactor has twenty-four coupons. These coupons can be made from a range of materials, from glass to polystyrene to metals. The biofilms formed in these systems are hardier in that they are more firmly attached to the coupon surface, as a result growing under conditions with shear. Both reactors provide replicate samples.

Flow-through systems use flow cells or tubing in which the media is pumped through the system [16]. Flow rates (fluid shear forces) can be more easily and precisely controlled in this system. The use of glass flow cells has shown to be a valuable instrument for real-time visualization of biofilms and treatment effects using the CSLM. This type of system can also be easily modified for testing of actual urinary catheters.

The drip flow reactor (DFR) is designed to model a low-shear environment and has been used for both single species and mixed species biofilm research [43]. The standard DFR model can hold four removable slides that serve as a substratum for biofilm attachment. We have observed that biofilms grown in a DFR tend to be thicker than those grown in other biofilm model systems. Presumably, bacteria within these biofilms have greater variations in physiologic and metabolic states. Some biofilms grown in this reactor have shown to be very resistant to treatments such as 10% bleach, with which there was only a 2 log reduction in bacterial counts after a 10 min treatment [44]. We have used both glass and hydroxy apatite (HA)–coated glass slides and have found that biofilms adhere better to the HA-coated glass surface. A variety of other materials can also be used as a substratum.

Problems with detection of biofilm on implanted devices

Bacteria attached to a medical device within the human body may escape detection by traditional clinical methodologies used to determine infection. Biofilm infections tend to remain subacute and chronic [45,46]. While the biofilm itself may contain very high cell numbers (10^4 to 10^8 cfu/cm^2), there may be no indication in the patient of such high population of bacteria because the bacteria are embedded in a matrix and because few bacteria are sloughing or shedding. In some cases, the patient may present with a low-grade infection. Patients suffering from catheter-related bloodstream infections often present with clinical symptoms of sepsis but with no apparent source of infection [47].

Blood culture is the standard method for the detection of infection (sepsis). Unfortunately, this method does not always detect an infection on an implanted medical device. Bacteria may be sloughed or shed from an established biofilm into the bloodstream, usually in much lower numbers than the total number of bacteria attached to the device surface. Those bacteria shed from the biofilm as single cells may be killed by host cells or antibiotics. In other words, the presence of a biofilm-infected implanted medical device can be easily overlooked since accepted clinical methods may not detect the presence of an infection. The patient may experience recurrent low-grade infections that may be temporarily eased by the administration of antibiotics. Or the patient may appear to be healthy for quite some time until there is an event that compromises the patient's immune system. If any compromise or suppression of the patient's immune system occurs, then the dissemination of the infection from the nidus of the biofilm-infected device may prove to be rapid and fatal.

Other potential tools for detecting biofilm infections

Environmental microbiology has led the way in the development of innovative tools to look for microorganisms that may be difficult to culture in the lab [48]. The use of molecular tools, including polymerase chain reaction (PCR) to amplify DNA from complex samples, denaturing gradient gel electrophoresis (DGGE) to assess patterns of primer-selected DNA, and BLAST analysis of DNA sequences, has shown a complexity in the microbial world that up until the molecular age was unknown. Researchers have applied these molecular techniques to the clinical setting in the successful search of causative agents for diseases for which no causative bacterial agent could be detected [49]. While not a biofilm infection, the discovery of the causative agent for Whipple's disease is a case in point. It was presumed that Whipple's disease was caused by a bacterial agent due to ancillary data such as patient response to antibiotic treatment. The use of molecular techniques showed that Whipple's was caused by a previously uncharacterized pathogen. Thus, the use of molecular techniques has provided valuable information regarding the microbial diversity of disease.

This illustrates that not all clinically relevant microorganisms are culturable despite the long history of Koch's postulates to successfully pinpoint causative agents of disease. Biofilms formed in the human body may be single species or a multispecies community. These biofilm communities are complex and usually go through a process of ecological succession, with certain species of microorganisms acting as the pioneer species, establishing the biofilm and later species inhabiting it. The complexity of dental plaque is an excellent example of how complex these biofilms can be, with an estimated 500+ species inhabiting the human oral cavity [50].

Microorganisms subjected to antibiotics may not always be killed, but instead may be injured. These bacteria may not be detectable under normal culture conditions but may be capable of resuscitation with time, and thus become a source of reinfection [51]. Even bacteria not exposed to antibiotics may be difficult to detect using culture methods. In a metabolic state, termed viable nonculturable (VNC), bacteria are alive but do not grow in culture. A study of urine specimens obtained from healthy mice showed the presence of bacteria in urine considered to be sterile [52]. The VNC state has been studied in bacteria in water and food-related environments, but little research has been done in clinically relevant settings. A number of studies of *Vibrio cholerae* have detected the presence of VNC strains in water [53] and have examined the presence of pathogens such as *Campylobacter* in chicken [54]. *Salmonella typhimurium* cells in sewage effluent were shown to enter a VNC state and retain their virulence after treatment with peracetic acid [55]. An important point concerning the VNC state of bacterial cells is that the effects of treatment may be overestimated because these cells are only capable of growth under certain conditions. Since bacteria in biofilms exist in different physiologic states, it is possible that VNC cells exist in clinically relevant biofilms.

Resistance to antimicrobials

It is well established that biofilms are resistant to antimicrobials [56]. It is important to note, however, that in most cases, the antimicrobial resistance seen in biofilms is not due to a genetic change. If the biofilm is disrupted, the bacterial cells are rendered susceptible to the antimicrobial agent. Depending on the species present and the antimicrobial used, this resistance may be due to different mechanisms. The matrix itself may provide a physical barrier to prevent the penetration of the antimicrobial through the biofilm. In other cases, however, it has been shown that certain antimicrobials can penetrate biofilm quickly and efficiently, but there is minimal effect of the antimicrobial on the cells within the biofilm. Bacterial cells within an established biofilm exist in different metabolic states. Cells at the surface of the biofilm experience an environment higher in nutrients and oxygen than cells contained deep within the matrix, where nutrients and oxygen may be severely limited. This leads to variations in metabolic and physiologic activities of the same species of bacteria within the biofilm, thereby affecting those mechanisms that certain antimicrobials may utilize for bactericidal effect. Bacteria not in an active metabolic and growth state, for example, are not susceptible to certain antibiotics, such as beta-lactams [57].

Evasion of host defenses

In addition to intrinsic resistance to antibiotics, biofilms have also been shown to be resistant to host defense mechanisms. A study characterizing the interaction of purified human neutrophils with *Pseudomonas aeruginosa* PAO1 biofilms showed that the neutrophils in contact with the biofilm became phagocytically engorged, partially degranulated, immobilized, and rounded, and showed only a slight increase in the soluble concentration of hydrogen peroxide [58].

Strains of *Serratia marcescens*, a common colonizer of contact lenses, have been shown to not induce a respiratory burst from polymorphonuclear leukocytes (PMNs) and to resist phagocytosis regardless of opsonization. This phenomenon was shown to increase when bacteria were grown as a biofilm on a contact lens [59].

It has been hypothesized that resistance to antimicrobial peptides may be due to interaction with biofilm and capsule exopolymers [60]. *Staphylococcus epidermidis* produces poly-N-acetylglucosamine (PIA) as a component of its exopolysaccharide during biofilm formation. PIA has been shown to play a crucial role in *S. epidermidis* resistance to neutrophil phagocytosis and human antibacterial peptides [61].

Device-associated biofilms

Any medical device can become colonized with bacteria. As the colonization process proceeds, biofilms can form in a relatively short period of time. Once established, biofilms are difficult to eradicate. The following discussion highlights some selected device-related biofilm infections.

Central venous catheters

There are approximately 5 million central venous catheters (CVCs) implanted yearly in the United States. Of those, it is estimated that 12 to 25% become infected, and 3 to 8% of patients with infections die [5]. Unfortunately, these numbers may well represent a gross underestimation of the problem. Reported numbers are usually only based on ICU infections and may not include patients using CVCs on an outpatient basis or in a home health care setting, such as for chemotherapy or for total parenteral nutrition. In addition, there are different types of vascular catheters (peripheral venous, central venous, arterial), and then the nomenclature becomes further complicated by duration of placement, insertion site (subclavian, femoral, etc.), the insertion pathway (tunneled, nontunneled), physical length, and other characteristics, such as the number of lumens or the presence of a cuff. Thus, the definition of the catheter may impact the report of infection numbers [5].

Catheter-related bloodstream infections (CRBSIs) are defined as the isolation of the same microorganisms from cultures of the distal catheter segment and patient blood cultures [47]. CRBSI is considered to be a major complication for patients with central venous catheters [62]. Rates for CRBSI are up to 13% of CVC and may be dependent on the catheter placement. Studies indicate that femoral catheters have a higher risk of infection in adults than other placement sites [63].

The sources of infection in a CVC are from the skin at the site of insertion, contamination of the device, spread of bacteria from another infection site through the blood (hematogenous), or infusion of contaminated fluid [5]. In addition, patients who test positive as nasal carriers for *S. aureus* have been shown to have a greater risk for CRBSI than patients who are noncarriers [5,64].

Urinary catheters

More than 30 million urinary catheters are placed yearly in the United States, with approximately 10 to 30% of those becoming infected, making them one of the most common causes of nosocomial infections. Fortunately, unlike CVCs, less than 5% of infections from urinary catheters result in death [65]. Visible biofilm containing countable bacterial numbers (10^4 cfu/cm^2) were detected in a patient catheter at 19 h postinsertion [66]. Despite the fact that the bladder is considered to be a sterile environment, the introduction of a urinary catheter can rapidly change this balance. Even if the perineum is swabbed before the catheter is inserted, it is only disinfected and not sterilized. As the catheter is inserted, it can drag bacteria from the outside skin surface to the bladder. The presence of the catheter also can trigger an inflammatory response and allows for pooling of urine in the bladder or catheter [65].

A number of strategies have been examined for the prevention of infection and biofilm formation on the catheter surface. Different types of materials, such a siliconized surfaces, and impregnating the surfaces with antimicrobial agents have been shown to be somewhat effective for patients undergoing short-term catheterization [65]. However, for patients with chronic indwelling catheters, such as the elderly or persons with spinal cord injuries, these strategies are not as successful. In most cases, these strategies merely delay the onset of biofilm formation.

Cerebrospinal fluid shunts

Cerebrospinal fluid (CSF) shunts are commonly used in the treatment of hydrocephalus to alleviate fluid pressure. There are approximately 70,000 hospital admissions for hydrocephalus, with the number of CSF shunt placements reaching tens of thousands annually [67,68]. Infection rates associated with CSF shunts are estimated to range from 3 to 30% [67]. An estimated 40% of the CSF shunts placed in pediatric patients fail within a year of implantation [68].

Sources of infection in CSF shunts may be related to the cause of the hydrocephalus, such as meningitis. However, in some cases, the placement of the shunt tip in the peritoneum has been implicated, especially in patients with appendicitis [67]. Thus, prevention of infection in CSF shunts is also complicated by the reason for the placement.

Scanning electron micrographs (SEMs) of CSF shunts removed from patients exhibiting symptoms of chronic infection indicate biofilm formation on the inner and outer surfaces of the shunts. In some cases, the presence of both cocci and bacilli indicates that the biofilms may be multispecies in nature [69].

Biofilms and diagnostic instruments

Cluster infections discovered in patients who are undergoing the same routine diagnostic exam at the same hospital or clinic are usually the first indicators of the presence of a biofilm in diagnostic instruments. The bacteria associated with these infections are resistant to routine disinfection procedures and are often associated with processing fluids, thus implicating biofilms. A variety of tests are usually performed to determine the infection source. These tests include routine swabbing and culture methods to find the source of bacterial contamination. In addition, molecular analyses are often performed to verify the strain of infecting bacterial species is the same in the affected patients as well as on the instruments in question.

Transrectal probes

Transrectal ultrasound-guided prostate biopsies are performed in men to diagnose prostate cancer. The procedure involves the use of an ultrasound probe to visualize the prostate, and then a needle biopsy is performed. In one case of cluster infections, the narrow lumen of the needle guide after it was removed from orthophthaldehyde (OPA) disinfectant was found to be culture positive for *P. aeruginosa*. The procedure to detect *P. aeruginosa* in the needle guide required scraping of the guide with a sterile needle, thus implicating the presence of a bacterial biofilm resistant to disinfection [70].

Sublingual probes

Sublingual probes are used to monitor tissue carbon dioxide levels in hospital patients. A cluster infection of *Burkholderia cepacia* was traced to the buffered saline solution in which the probes are packaged. Further testing of the saline revealed not only *B. cepacia*, but other gram-negative rods, thus implicating the processing water at the manufacturer [71].

Endoscopes

Endoscopic procedures are minimally invasive diagnostic techniques that include a range of procedures, such as colonoscopies, laproscopies, and rhinoscopies. Cluster infections as a result of endoscopic procedures have been reported and have led to septicemia and cholangitis or infection of the bile ducts. Bacterial species detected include *P. aeruginosa*, *Mycobacterium tuberculosis*, *M. intracellulare*, and *M. avium-intracellulare* [72,73].

The manual cleaning of endoscopes is laborious due to the design of the endoscope, which contains small lumens, relatively long tubes, and angles. In addition, standard disinfection methods recommend the use of glutaraldehyde, and thereby can present a chemical risk hazard to health care personnel in charge of manually cleaning the scopes. Often, then, endoscopes are routinely cleaned and reprocessed in automated machines specially designed for that purpose. These machines use a detergent solution rinse, a disinfection step that may include glutaraldehyde, phenol, or some combination of disinfectants, and a final rinse, which is often performed with tap water. Biofilms have been found in these reprocessing machines in the detergent-holding tanks, hoses, and air vents [72,73]. Recently, biofilm-targeted approaches have been shown to provide better disinfection and cleaning than standard protocols [74].

The bottom line

A common theme for the prevention of biofilm formation on medical devices includes prevention of device contamination. There must be rigorous attention paid to aseptic technique in surgery, clinics, and home health care situations. Hand washing is the most common procedure described in the prevention of cross-contamination [2]. For infected devices, the best option is often to remove the device as soon as possible. The use of tap water, which is not sterile, for the rinsing or processing of medical devices should be strongly discouraged.

At this point in time, there is no magic bullet that prevents the formation of biofilms on surfaces. Once bacteria become established as a biofilm on a medical device, the prognosis becomes somewhat grim, especially in the case of immune-compromised or -suppressed patients. For the present, the best protocol is the prevention of contamination of the device in order to prevent those contaminating bacteria from attaching to a surface and developing into a biofilm.

References

1. Graves, N., Economics and preventing hospital-acquired infection, *Emerg Infect Dis*, 10, 561, 2004.
2. McCaughey, B., Unnecessary deaths: *The human and financial costs of hospital infections*, Committee to Reduce Infection Deaths, www.hospitalinfection .org, 2005.
3. Wenzel, R.P., Edmond, M.B., The impact of hospital-acquired bloodstream infections, *Emerg Infect Dis*, 7, 174, 2001.
4. Roberts, R.R., Scott, R.D., 2nd, Cordell, R., Solomon, S.L., Steele, L., Kampe, L.M., Trick, W.E., Weinstein, R.A., The use of economic modeling to determine the hospital costs associated with nosocomial infections, *Clin Infect Dis*, 36, 1424, 2003.
5. O'Grady, N.P., Alexander, M., Dellinger, E.P., Gerberding, J.L., Heard, S.O., Maki, D.G., Masur, H., McCormick, R.D., Mermel, L.A., Pearson, M.L., Raad, I.I., Randolph, A., Weinstein, R.A., Guidelines for the prevention of intravascular catheter-related infections, *MMWR Recomm Rep*, 51, 1, 2002.
6. Guadagnino, C., Pennsylvania's hospital-acquired infection battle, *Physician's Digest*, www.physiciansnews.com/cover/206.html, 2006.
7. Braxton, E.E., Jr., Ehrlich, G.D., Hall-Stoodley, L., Stoodley, P., Veeh, R., Fux, C., Hu, F.Z., Quigley, M., Post, J.C., Role of biofilms in neurosurgical device-related infections, *Neurosurg Rev*, 28, 249, 2005.
8. Uslan, D.Z., Sohail, M.R., Friedman, P.A., Hayes, D.L., Wilson, W.R., Steckelberg, J.M., Baddour, L.M., Frequency of permanent pacemaker or implantable cardioverter-defibrillator infection in patients with gram-negative bacteremia, *Clin Infect Dis*, 43, 731, 2006.
9. Caiazza, N.C., O'Toole, G.A., SadB is required for the transition from reversible to irreversible attachment during biofilm formation by *Pseudomonas aeruginosa* PA14, *J Bacteriol*, 186, 4476, 2004.
10. de Lancey-Pulcini, E., Effects of initial adhesion events on the physiology of *Pseudomonas aeruginosa*, PhD dissertation, Montana State University, 2001.
11. Ghigo, J.M., Are there biofilm-specific physiological pathways beyond a reasonable doubt? *Res Microbiol*, 154, 1, 2003.
12. O'Toole, G., Kaplan, H.B., Kolter, R., Biofilm formation as microbial development, *Annu Rev Microbiol*, 54, 49, 2000.

13. Petrelli, D., Zampaloni, C., D'Ercole, S., Prenna, M., Ballarini, P., Ripa, S., Vitali, L.A., Analysis of different genetic traits and their association with biofilm formation in *Staphylococcus epidermidis* isolates from central venous catheter infections, *Eur J Clin Microbiol Infect Dis*, 25, 773, 2006.
14. de Lancey-Pulcini, E., James, G., unpublished data, 2006.
15. Fux, C.A., Wilson, S., Stoodley, P., Detachment characteristics and oxacillin resistance of *Staphyloccocus aureus* biofilm emboli in an in vitro catheter infection model, *J Bacteriol*, 186, 4486, 2004.
16. Rupp, C.J., Fux, C.A., Stoodley, P., Viscoelasticity of *Staphylococcus aureus* biofilms in response to fluid shear allows resistance to detachment and facilitates rolling migration, *Appl Environ Microbiol*, 71, 2175, 2005.
17. Costerton, J.W., Montanaro, L., Arciola, C.R., Biofilm in implant infections: Its production and regulation, *Int J Artif Organs*, 28, 1062, 2005.
18. Hall-Stoodley, L., Brun, O.S., Polshyna, G., Barker, L.P., *Mycobacterium marinum* biofilm formation reveals cording morphology, *FEMS Microbiol Lett*, 257, 43, 2006.
19. Heyrman, J., Verbeeren, J., Schumann, P., Swings, J., De Vos, P., Six novel *Arthrobacter* species isolated from deteriorated mural paintings, *Int J Syst Evol Microbiol*, 55, 1457, 2005.
20. Oosthuizen, M.C., Steyn, B., Theron, J., Cosette, P., Lindsay, D., Von Holy, A., Brozel, V.S., Proteomic analysis reveals differential protein expression by *Bacillus cereus* during biofilm formation, *Appl Environ Microbiol*, 68, 2770, 2002.
21. Sanchez-Moral, S., Luque, L., Cuezva, S., Soler, V., Benavente, D., Laiz, L., Gonzalez, J.M., Saiz-Jimenez, C., Deterioration of building materials in Roman catacombs: The influence of visitors, *Sci Total Environ*, 349, 260, 2005.
22. Steinberg, D., Rozen, R., Bromshteym, M., Zaks, B., Gedalia, I., Bachrach, G., Regulation of fructosyltransferase activity by carbohydrates, in solution and immobilized on hydroxyapatite surfaces, *Carbohydr Res*, 337, 701, 2002.
23. Teughels, W., Van Assche, N., Sliepen, I., Quirynen, M., Effect of material characteristics and/or surface topography on biofilm development, *Clin Oral Implants Res*, 17, 68, 2006.
24. Verran, J., Whitehead, K., Factors affecting microbial adhesion to stainless steel and other materials used in medical devices, *Int J Artif Organs*, 28, 1138, 2005.
25. Wingender, J., Flemming, H.C., Contamination potential of drinking water distribution network biofilms, *Water Sci Technol*, 49, 277, 2004.
26. Cerca, N., Pier, G.B., Vilanova, M., Oliveira, R., Azeredo, J., Influence of batch or fed-batch growth on *Staphylococcus epidermidis* biofilm formation, *Lett Appl Microbiol*, 39, 420, 2004.
27. de Lancey-Pulcini, E., James, G., unpublished data, 2006.
28. Carlen, A., Rudiger, S.G., Loggner, I., Olsson, J., Bacteria-binding plasma proteins in pellicles formed on hydroxyapatite in vitro and on teeth in vivo, *Oral Microbiol Immunol*, 18, 203, 2003.
29. Uyen, H.M., Schakenraad, J.M., Sjollema, J., Noordmans, J., Jongebloed, W.L., Stokroos, I., Busscher, H.J., Amount and surface structure of albumin adsorbed to solid substrata with different wettabilities in a parallel plate flow cell, *J Biomed Mater Res*, 24, 1599, 1990.

30. Bruinsma, G.M., van der Mei, H.C., Busscher, H.J., Bacterial adhesion to surface hydrophilic and hydrophobic contact lenses, *Biomaterials*, 22, 3217, 2001.
31. Stickler, D.J., Jones, S.M., Adusei, G.O., Waters, M.G., A sensor to detect the early stages in the development of crystalline *Proteus mirabilis* biofilm on indwelling bladder catheters, *J Clin Microbiol*, 44, 1540, 2006.
32. Sutherland, I., Biofilm exopolysaccharides: A strong and sticky framework, *Microbiology*, 147, 3, 2001.
33. Klapper, I., Dockery, J., Role of cohesion in the material description of biofilms, *Phys Rev E Stat Nonlin Soft Matter Phys*, 74, 031902, 2006.
34. Somerville, G.A., Beres, S.B., Fitzgerald, J.R., DeLeo, F.R., Cole, R.L., Hoff, J.S., Musser, J.M., In vitro serial passage of *Staphylococcus aureus*: Changes in physiology, virulence factor production, and agr nucleotide sequence, *J Bacteriol*, 184, 1430, 2002.
35. Nilsson, A.I., Kugelberg, E., Berg, O.G., Andersson, D.I., Experimental adaptation of *Salmonella typhimurium* to mice, *Genetics*, 168, 1119, 2004.
36. Boles, B.R., Thoendel, M., Singh, P.K., Self-generated diversity produces "insurance effects" in biofilm communities, *Proc Natl Acad Sci USA*, 101, 16630, 2004.
37. Haussler, S., Biofilm formation by the small colony variant phenotype of *Pseudomonas aeruginosa*, *Environ Microbiol*, 6, 546, 2004.
38. Pitts, B., Hamilton, M.A., Zelver, N., Stewart, P.S., A microtiter-plate screening method for biofilm disinfection and removal, *J Microbiol Methods*, 54, 269, 2003.
39. O'Toole, G.A., and Kolter, R., Initiation of biofilm formation in *Pseudomonas fluorescens* WCS365 proceeds via multiple, convergent signalling pathways: A genetic analysis. *Mol Microbiol*, 28, 449, 1998.
40. Harrison, J.J., Turner, R.J., Ceri, H., High-throughput metal susceptibility testing of microbial biofilms, *BMC Microbiol*, 5, 53, 2005.
41. Goeres, D.M., Loetterle, L.R., Hamilton, M.A., Murga, R., Kirby, D.W., Donlan, R.M., Statistical assessment of a laboratory method for growing biofilms, *Microbiology*, 151, 757, 2005.
42. Zelver, N., Hamilton, M., Goeres, D., Heersink, J., Development of a standardized antibiofilm test, *Methods Enzymol*, 337, 363, 2001.
43. Adams, H., Winston, M.T., Heersink, J., Buckingham-Meyer, K.A., Costerton, J.W., Stoodley, P., Development of a laboratory model to assess the removal of biofilm from interproximal spaces by powered tooth brushing, *Am J Dent*, 15, 12B, 2002.
44. de Lancey-Pulcini, E., James, G., unpublished data, 2006.
45. Thomas, C., Cadwallader, H.L., Riley, T.V., Surgical-site infections after orthopaedic surgery: Statewide surveillance using linked administrative databases, *J Hosp Infect*, 57, 25, 2004.
46. Ehrlich, G.D., Stoodley, P., Kathju, S., Zhao, Y., McLeod, B.R., Balaban, N., Hu, F.Z., Sotereanos, N.G., Costerton, J.W., Stewart, P.S., Post, J.C., Lin, Q., Engineering approaches for the detection and control of orthopaedic biofilm infections, *Clin Orthop Relat Res*, 437, 59, 2005.
47. Gastmeier, P., Geffers, C., Prevention of catheter-related bloodstream infections: Analysis of studies published between 2002 and 2005, *J Hosp Infect*, 64, 326, 2006.

48. Amman, R.I., Ludwig, W., Schleifer, K.H., Phylogenetic identification and in situ detection of individual microbial cells without cultivation, *Microbiol Rev*, 59, 143, 1995.
49. Maiwald, M., Relman, D., Whipple's disease and *Tropheryma whippelii*: Secrets slowly revealed, *Clin Infect Dis*, 32, 457, 2001.
50. Kolenbrander, P.E., Oral microbial communities: Biofilms, interactions, and genetic systems, *Annu Rev Microbiol*, 54, 413, 2000.
51. Arrese, J.E., Goffin, V., Avila-Camacho, M., Greimers, R., Pierard, G.E., A pilot study on bacterial viability in acne. Assessment using dual flow cytometry on microbials present in follicular casts and comedones, *Int J Dermatol*, 37, 461, 1998.
52. Anderson, M., Bollinger, D., Hagler, A., Hartwell, H., Rivers, B., Ward, K., Steck, T.R., Viable but nonculturable bacteria are present in mouse and human urine specimens, *J Clin Microbiol*, 42, 753, 2004.
53. Binsztein, N., Costagliola, M.C., Pichel, M., Jurquiza, V., Ramirez, F.C., Akselman, R., Vacchino, M., Huq, A., Colwell, R., Viable but nonculturable *Vibrio cholerae* O1 in the aquatic environment of Argentina, *Appl Environ Microbiol*, 70, 7481, 2004.
54. Wolffs, P., Norling, B., Hoorfar, J., Griffiths, M., Radstrom, P., Quantification of *Campylobacter* spp. in chicken rinse samples by using flotation prior to real-time PCR, *Appl Environ Microbiol*, 71, 5759, 2005.
55. Jolivet-Gougeon, A., Sauvager, F., Bonnaure-Mallet, M., Colwell, R.R., Cormier, M., Virulence of viable but nonculturable *S. Typhimurium* LT2 after peracetic acid treatment, *Int J Food Microbiol*, 112, 147, 2006.
56. Stewart, P.S., Mechanisms of antibiotic resistance in bacterial biofilms, *Int J Med Microbiol*, 292, 107, 2002.
57. Anderl, J.N., Zahller, J., Roe, F., Stewart, P.S., Role of nutrient limitation and stationary-phase existence in *Klebsiella pneumoniae* biofilm resistance to ampicillin and ciprofloxacin, *Antimicrob Agents Chemother*, 47, 1251, 2003.
58. Jesaitis, A.J., Franklin, M.J., Berglund, D., Sasaki, M., Lord, C.I., Bleazard, J.B., Duffy, J.E., Beyenal, H., Lewandowski, Z., Compromised host defense on *Pseudomonas aeruginosa* biofilms: Characterization of neutrophil and biofilm interactions, *J Immunol*, 4329, 2002.
59. Hume, E.B., Stapleton, F., Willcox, M.D., Evasion of cellular ocular defenses by contact lens isolates of *Serratia marcescens*, *Eye Contact Lens*, 29, 108, 2003.
60. Otto, M., Bacterial evasion of antimicrobial peptides by biofilm formation, *Curr Top Microbiol Immunol*, 306, 251, 2006.
61. Vuong, C., Kocianova, S., Voyich, J.M., Yao, Y., Fischer, E.R., DeLeo, F.R., Otto, M., A crucial role for exopolysaccharide modification in bacterial biofilm formation, immune evasion, and virulence, *J Biol Chem*, 280, 12064, 2005.
62. Bacuzzi, A., Cecchin, A., Del Bosco, A., Cantone, G., Cuffari, S., Recommendations and reports about central venous catheter-related infection, *Surg Infect*, 7, S65, 2006.
63. Lorente, L., Santacreu, R., Martin, M.M., Jimenez, A., Mora, M.L., Arterial catheter-related infection of 2,949 catheters, *Crit Care*, 10, R83, 2006.
64. von Eiff, C., Becker, K., Machka, K., Stammer, H., Peters, G., Nasal carriage as a source of *Staphylococcus aureus* bacteremia, *N Engl J Med*, 344, 11, 2001.
65. Trautner, B.W., Darouiche, R.O., Role of biofilm in catheter-associated urinary tract infection, *Am J Infect Control*, 32, 177, 2004.

66. de Lancey-Pulcini, E., James, G., unpublished data, 2006.
67. Fux, C.A., Quigley, M., Worel, A.M., Post, C., Zimmerli, S., Ehrlich, G., Veeh, R.H., Biofilm-related infections of cerebrospinal fluid shunts, *Clin Microbiol Infect*, 12, 331, 2006.
68. Kulkarni, A.V., Drake, J.M., Lamberti-Pasculli, M., Cerebrospinal fluid shunt infection: A prospective study of risk factors, *J Neurosurg*, 94, 195, 2001.
69. Drake, J.M., et al., Randomized trial of cerebrospinal fluid shunt valve design in pediatric hydrocephalus, *Neurosurgery*, 43, 294, 1998.
70. CDC, *Pseudomonas aeruginosa* infections associated with transrectal ultrasound-guided prostate biopsias—Georgia, 2005, *MMWR*, 55, 776, 2006.
71. CDC, Nosocomial *Burkholderia cepacia* infections associated with exposure to sublingual probes—Texas, 2004, *MMWR* 53, 796, 2004.
72. CDC, Bronchoscopy-related infections and pseudoinfections—New York, 1996 and 1998, *MMWR*, 48, 557, 1999.
73. CDC, Nosocomial infection and pseudoinfection from contaminated endoscopes and bronchoscopes—Wisconsin and Missouri, *MMWR*, 40, 675, 1991.
74. Marion, K., Freney, J., James, G., Bergeron, E., Renaud, F.N., Costerton, J.W., Using an efficient biofilm detaching agent: An essential step for the improvement of endoscope reprocessing protocols, *J Hosp Infect*, 64, 136, 2006.

chapter six

A simulated-use evaluation of a strategy for preventing biofilm formation in dental unit waterlines*

James W. McDowell, Daryl S. Paulson, and John A. Mitchell

Contents

Dental unit waterline contamination can pose a serious threat of infection to patients. Such contamination is the result of biofilms that adhere to the inner surfaces of the lines. Biofilms consist of bacterial cells immobilized in an organic polymer matrix that often is highly resistant to removal. The biofilm protects the bacteria both from being washed away by the water flow and from many types of antimicrobial water treatment [1,2].

In 1995, the American Dental Association (ADA) Board of Trustees adopted a statement on dental unit waterlines directing industry and the research community to take an "ambitious and aggressive course" to ensure the delivery of quality treatment water to patients [3]. Before the statement's publication, research findings revealing unacceptable bacterial contamination in dental unit water were becoming common [4–7]. In subsequent years, the dental profession developed a common understanding that dental unit waterlines will become contaminated if

---

no control measures are applied [8]; as a consequence, during that time numerous products were introduced that addressed the dental unit waterline biofilm issue with some measure of success. Beyond the concern for efficacy, however, most of these products have issues of material incompatibility with dental equipment and are difficult or costly for the practitioner to use [9].

Methods intended to improve the microbiological quality of dental unit water can be classified as preventive or remedial. For instance, microfiltration of the outflow of a dental unit water system is a remedial method. With disregard to the contamination upstream, these filters are designed to "catch" contamination before exposing the patient. On the contrary, however, the ADA statement indicated that bacteria should be controlled in the "unfiltered output" of the dental unit because the association recognized the need for systemic control of bacteria throughout the dental unit water system, not just at one point, such as that provided by a filter [7].

Another remedial method commonly used for controlling the quality of dental unit water is periodic shock treatment of the waterlines with an aggressive chemical, such as sodium hypochlorite or certain commercial products distributed for this purpose. Because such shock treatments are intended to eliminate contamination once it has developed on the internal surfaces of waterlines, their cleanup approach is a remedial method. The success of these treatment regimens is highly dependent on the quality of the water that is supplied daily to the system, because there typically is no residual protection against deterioration of water quality between applications. In addition, they generally are not convenient to apply, and their effectiveness often depends on administrative compliance measures. Furthermore, the shocking process, by its nature, is chemically aggressive, thereby resulting in increased potential for damage to the dental unit water system.

Ideally, a treatment process would prevent the development of biofilm contamination of the water system, be easily performed, and offer continuous protection, thereby eliminating the root cause of poor dental unit water quality. Also, the process should provide continued efficacy during periods of nonuse, such as overnight and weekends. A treatment process exhibiting these attributes would be much easier to explain to patients and easier for practitioners to manage than remedial treatment processes.

A proactive approach is taken by a new product, A-Dec ICX waterline tablets (A-Dec, Newburg, Oregon), an effervescent tablet that is added directly to the dental unit water bottle at each refill. The product contains multiple active ingredients, including sodium percarbonate, silver nitrate, and cationic surfactants, to provide both immediate and sustained residual protection against biofilm formation. The ingredients in

ICX are regarded as safe for human consumption based on the incidental ingestion model accepted by the U.S. Food and Drug Administration for premarket clearance, consistent with generally recognized as safe ingestion models.

In this chapter, we describe our evaluation of the ICX system for its ability to prevent biofilm formation. We addressed several research questions in this study:

- How effective is the test product in preventing biofilm/microbial growth in dental unit waterlines during an extended period of simulated use?
- Are there any significant differences in treatment effects between the hand-piece water coolant (HPWC) lines and the air-water syringe (AWS) line?
- Does the test product maintain acceptable water quality throughout the week and after weekends?
- Is the test product's effectiveness affected by water hardness?

This study simulated conditions in dental unit waterlines over a period of approximately 4 months. Our basic study design involved creating conditions that promoted growth of biofilms in dental unit water systems, using bacteria common in dental unit water and the environment. The test product was used to treat the test units, and effluent bacterial counts were tracked in both test and control (untreated) test units. In addition, we used scanning electron microscopy (SEM) to inspect internal tubing surfaces for biofilm formation.

An added objective of this study was the validation of a laboratory test method for reproducibly assessing the efficacy of proposed treatment processes. Other published studies attempting to replicate dental unit biofilm formation in the laboratory generally have described inadequate models that are not well suited to evaluating preventive treatment approaches. The test apparatus and procedures used in this study accurately reproduced the dental unit water flow typically observed with use of a dental unit.

Materials and methods

To re-create dental unit waterline conditions accurately and precisely over time, we used a series of ten automated dental unit water systems (test units). Each of the ten independent test units contained all of the components of a typical water delivery system, which included a water bottle, polyurethane supply tubing, a control system, three high-speed HPWC lines, and one AWS line. (We used A-Dec 500 (A-Dec) components in the test units.) An electronic controller operated the test unit waterlines intermittently to simulate daily dental unit usage, using approximately 60 ml per simulated patient, a volume consistent with published clinical data [10]. The test program consisted of ten simulated patient treatment cycles per day and a flushing of all waterlines at the start of each day and after each patient, as recommended by the Centers for Disease Control and Prevention [11]. We added the test product, A-Dec ICX, to the water reservoir bottle at each refill. The test product does not contain any chlorine compounds associated with the production of trihalomethanes, a class of compounds that includes a number of suspected carcinogens [12].

We divided the ten test units into two groups of five, treating three of the test units in each group with the test product and using the remaining two as the untreated controls. In one group of five units, we filled the water bottles with distilled water buffered with a 1:100 dilution of phosphate-buffered saline. In the other group, we filled the bottles with identically buffered distilled water in which water hardness was adjusted to at least 200 mg/L as calcium carbonate, or $CaCO_3$. We selected the hardness level as being representative of typical U.S. hard municipal water, based on a citation that 94% of the one hundred largest cities in the United States were found to have a water hardness of less than 200 mg/l as $CaCO_3$ [13]. We will refer to these solutions as buffered distilled water and hard water, respectively. We inoculated the solutions used to fill the water bottles of the test units with a pooled challenge suspension of *Klebsiella pneumoniae* (American Type Culture Collection, or ATCC, no. 4352), *Pseudomonas aeruginosa* (ATCC no. 15442), and *Staphylococcus aureus* (ATCC no. 6538) to a level of approximately 10^2 to 10^3 colony-forming units, or CFU, per milliliter. This inoculum level was shown to result reliably in the development of a biofilm (J.W. McDowell, D.S. Paulson, J.A. Mitchell, unpublished data, July 2003) and approximates the 500 CFU/m; maximum heterotrophic plate count allowable by the U.S. Environmental Protection Agency's National Primary Drinking Water Regulations. We then secured the inoculated bottles on the test units and pressurized the test units for the daily (5-days-per-week) controller cycle.

We performed bacterial sampling from the HPWC lines of each of the test units on the first working day of each week before we refilled the water bottles. We again sampled the test unit HPWC lines after daily refilling and operation of the controller cycle for at least 15 min. Finally, on the last working day of the week, we sampled the HPWC line and the AWS line after 4 h of controller cycle operation. Before doing any sampling, we disinfected the outer surfaces of the HPWC and AWS water outlets by swabbing them with 70% isopropyl alcohol, air-drying them for at least 30 s, and flushing them for at least 2 s. For each test unit, we sampled each of the three HPWC lines, and we pooled equal volumes from each for a composite HPWC sample. We plated the single weekly sample from the AWS line separately. We implemented a reduced sampling sequence comprising only composite HPWC lines before preparation on the first day of each week and after 4 h of operation on the last working day for weeks 13, 14, and 15 of the 16-week study.

We prepared all samples for analysis using a validated neutralization process that included sodium thiosulfate and sodium thioglycolate in the diluting and plating media, to eliminate any remaining antimicrobial activity. We performed serial tenfold dilutions and spread-plated them on R2A agar with neutralizers. We inverted the plates and incubated them at 20 to 25°C. After incubation, we enumerated colonies on the plates and characterized the microorganisms on the basis of morphology either as an expected type of the challenge species or as exogenous to the system.

At the conclusion of weeks 10 and 16, we randomly selected samples of waterline tubing, two HPWC lines and one AWS line, from each test group (that is, the untreated control group and treated group for each water type) and excised them for SEM analysis. We purged the selected waterlines with air before aseptically removing a 2 to 3 cm section 50 cm from the outlet end of the HPWC or AWS line. We then spliced the lines together for continued operation. We fixed and stored the SEM samples in a 2.5% glutaraldehyde solution. In preparation for sputter coating, we removed the samples from the fixing solution, rinsed them in water, sliced them longitudinally, and allowed them to air-dry for no more than 2 h. We performed sputter coating with gold/palladium and SEM analysis at the Image and Chemical Analysis Laboratory at Montana State University in Bozeman on the samples from weeks 10 and 16.

Results

The populations of the daily inoculum—a mean population of 198 CFU/ml (2.30 $\log_{10}$/ml) for the buffered distilled water test group and a mean population of 165 CFU/ml (2.22 $\log_{10}$/ml) for the hard water test group— were consistent over the course of the study. The level of water hardness was maintained within the range of 200 to 252 mg/L as $CaCO_3$ during the course of the study.

Biofilm development in the untreated control test units, regardless of water type, produced effluent counts that rapidly exceeded those of the inoculum population. Three of the four untreated controls achieved greater than 4.0 $\log_{10}$/ml by the beginning of the third week, while the fourth untreated control did not reach greater than 4.0 $\log_{10}$/ml until the eighth week. However, the lagging untreated control had exogenous microorganisms in the effluent water samples that were excluded from the total plate count until the ninth week.

In contrast to the control group results, water samples from the six treated test units showed no appreciable bacterial counts—less than 3 CFU/ml—during the 16-week period of study. Figures 6.1 and 6.2 show the

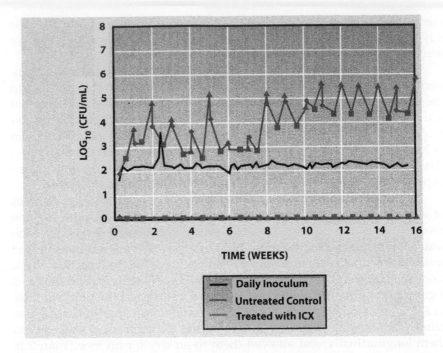

Figure 6.1 Bacteria levels in test units supplied with inoculated buffered distilled water over 16 weeks. For reference, the daily inoculation level also is shown. The untreated control group became contaminated rapidly, while the test product reduced the incoming bacteria levels to near zero for the duration of the study. CFU/ml, Colony-forming units per milliliter; ICX, A-Dec ICX waterline tablets (A-Dec, Newberg, Oregon). (Copyright© 2004 American Dental Association. All rights reserved. With permission.)

progression of mean plate counts in the effluent water from the test groups using buffered distilled water and hard water, respectively. For reference, the daily inoculum level also is shown. There is a 1-day gap in the data plotted for the treated HPWC sample before daily use at day 14 in both Figures 6.1 and 6.2, because no zero-dilution plates were prepared on that day.

We performed a series of two-sample Student t tests ($p = 0.05$) on data from samples taken during weeks 9 through 16, and we generated 95% confidence intervals. Because the control units supplied with hard water produced significantly higher bacterial populations than did those supplied with buffered distilled water, we evaluated the two groups of five test units separately. Regardless of group, however, comparative statistical analyses yielded the same outcomes. In both groups, the untreated control units produced significantly greater populations ($p < 0.05$) of bacteria than did treated units, which, in fact, were essentially free of viable bacteria. Among the treated units of both groups, there were no significant differences

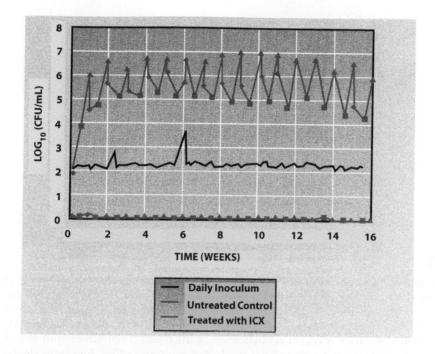

Figure 6.2 Plate count results for the test units supplied with hard water. For reference, the daily inoculation level also is shown. The results are similar to those for the buffered distilled water group, except that the level of contamination in the untreated control group was greater with hard water. CFU/ml, Colony-forming units per milliliter; ICX: A-Dec ICX waterline tablets (A-Dec, Newberg, Oregon). (Copyright© 2004 American Dental Association. All rights reserved. With permission.)

between bacterial populations attributable to variables such as the weekday on which or time of day at which samples were taken, or whether samples were from HPWC or AWS lines. Among samples from the untreated control units of both groups, populations did not differ significantly on the basis of source (HPWC or AWS line), but did differ on the basis of time variables. That is, populations in HPWC samples taken before the refilling of the bottles on the first day of the week (Monday after a weekend) were significantly greater ($p < .05$) than populations in samples taken 15 min after recharging, which, in turn, were significantly greater ($p < .05$) than populations in the 4 h sample taken on the last day of the week (Friday).

SEM analysis corroborated the plate count results. In both the week 10 and week 16 SEM images, a proliferation of bacterial colonization was apparent in all of the untreated test unit samples, while we observed no bacterial colonization in any of the specimens from the treated test units. Figures 6.3 and 6.4 are representative SEM images of samples from the treated and untreated test units, respectively.

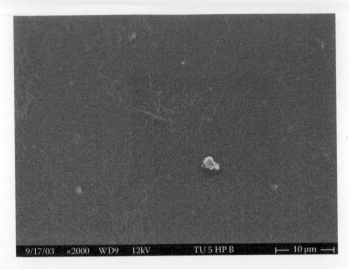

9/17/03   ×2000   WD9   12kV   TU 5 HP B   ⊢— 10 μm —⊣

Figure 6.3 This scanning electron micrograph (× 2,000 magnification) is typical of the surface of a handpiece water coolant line after 10 weeks of daily exposure to buffered distilled water inoculated with 10^2 to 10^3 colony-forming units per milliliter and treated with A-Dec ICX waterline tablets (A-Dec, Newberg, Oregon). The SEM analysis revealed no indication of microbial contamination in any of the test units treated with the test product. (Copyright© 2004 American Dental Association. All rights reserved. With permission.)

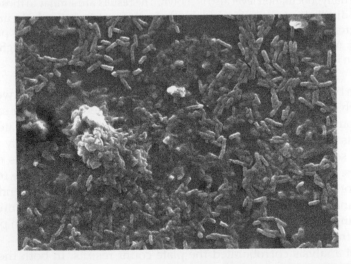

Figure 6.4 This scanning electron micrograph (× 2,000 magnification) is typical of the surface of a handpiece water coolant line after 10 weeks of daily exposure to buffered distilled water inoculated with 10^2 to 10^3 colony-forming units per milliliter and not treated with the test product (that is, a waterline from the control group). Numerous microorganisms in an established biofilm matrix are evident. (Copyright© 2004 American Dental Association. All rights reserved. With permission.)

Discussion

Considerable attention was directed toward making the test system representative of actual dental units in clinical service. Each test unit accurately recreated a dental unit water system. We modeled the pattern of use after typical daily clinical use, and we selected clinically relevant bacterial challenge species. One simplification in this laboratory test design to be noted is that we introduced the challenge species exclusively in the water added to the supply bottle, while in the clinical setting, there exists at least a theoretical possibility that contamination also could occur at the outlets of the waterlines. Sterilization of hand pieces and syringe tips subsequent to each patient treatment, use of nonretracting dental units, and compliance with CDC recommendations to flush after each patient [11] all are measures that substantially reduce this risk in the clinical setting.

We chose the challenge species to represent gram-positive (*S. aureus*) and gram-negative (*K. pneumoniae* and *P. aeruginosa*) microorganisms common in the environment. *S. aureus* is associated with skin and mucous membranes and is isolated from food products, dust, and water, including dental unit water [14]. *P. aeruginosa* and *K. pneumoniae* are microorganisms that produce glycocalices involved in biofilm formation and, reportedly, are common biofilm contaminants in dental unit water systems [5,14,15]. We added a dilute buffer to the challenge water to minimize the possibility of incidentally lysing the challenge bacteria owing to changes in osmotic pressure.

The water used to create the buffered distilled water and hard water was nonsterilized steam-distilled water. This resulted in the introduction of exogenous bacteria to all test units and a more diverse colonization of the untreated control test units in the buffered distilled water group. We identified *Brevundimonas vesicularis* and *Sphingomonas paucimobilis* on the basis of colony morphologies, but our analysis for exogenous species was not exhaustive.

We considered the study's first 8 weeks to be a conditioning period during which test unit performance was stabilized. We conducted statistical analyses using the total plate counts from all units from the ninth through sixteenth weeks. The data support consideration of a shorter conditioning period, but because we did not include exogenous organisms—which eventually composed a significant portion of the effluent counts from some of the test units—in plate count enumeration prior to week 9, we performed statistical analyses on data accumulated from that point. The spread plate results from treated test units consistently were less than 3 CFU/ml, demonstrating that the test product was capable of reducing the level of contamination in the daily challenge water and maintaining high water quality throughout the dental unit at all times.

After 16 weeks, there was still no indication that a breakthrough in or escalation of effluent plate counts might be forthcoming (Figures 6.1 and 6.2). This finding was reinforced by an SEM analysis that revealed no evidence of biofilm formation or colonization by any microorganisms in the treated test units (Figure 6.3). On the other hand, the effluent counts from the untreated test units clearly portrayed a developing biofilm, which was plainly revealed by the SEM analysis (Figure 6.4). Visual examinations and SEM analyses did not suggest any corrosion of or material damage to the treated test units.

Statistical analyses of the plate count data revealed significant differences between bacterial populations from the treated and untreated test units for all waterlines, times of day, and weekdays sampled. Populations in samples from the untreated test units declined with increasing time of operation through the day and the week. At no time, however, did flushing alone reduce the effluent count to the level of less than 200 CFU/ml recommended by the ADA [1]. In fact, populations were often a hundred to a thousand times greater than this recommended level. The plate counts observed in the untreated control units were consistent with the range reported in many studies on the microbiological water quality of dental units in actual clinical service [3,5,15,17].

Conclusion

In the presence of a bacterial challenge of 100 to 1,000 CFU/ml in the incoming water, A-Dec ICX effectively prevented the development of biofilm and maintained water quality at a level consistently well below 200 CFU/ml at both high and low water hardness levels under conditions that replicate clinical use. Persistence of the inhibition was observed during periods of inactivity typical of clinical practice. During the 16-week course of study, there was no breakthrough of microorganisms in the effluent samples from treated units, nor did we observe adherent microorganisms in SEM analyses of excised waterlines. On the other hand, the untreated controls developed extensive biofilm, resulting in contaminated water at levels consistent with clinical findings. This study did not seek to evaluate whether daily use of A-Dec ICX might reduce or eliminate the importance of flushing after each patient, as recommended by the CDC. Further study examining this question may be of interest.

This research demonstrates that A-Dec ICX effectively controlled bacterial contamination in dental unit waterlines and prevented biofilm buildup during daily use and over weekend periods of inactivity, thereby meeting the ADA's recommended goal for dental unit water quality. Following CDC recommendations and dispensing one tablet into the supply water bottle before daily operation achieved proactive biofilm prevention in dental unit water supplied in this study. While more studies are warranted to continue investigation of the product's efficacy in a variety of end-user conditions, this study provides evidence that A-Dec ICX offers dental practitioners promise for a convenient and effective way to maintain clean dental unit waterlines.

Disclosure

This study was supported by A-Dec, Newburg, Oregon, manufacturer of the biofilm prevention product described in this article.

Dr. Paulson is president and chief executive officer, Bioscience Laboratories, Bozeman, Montana.

Dr. Mitchell is director of quality assurance, Bioscience Laboratories, Bozeman, Montana.

References

1. Forbes BA, Sahm DF, Weissfeld AS. *Bailey & Scott's diagnostic microbiology,* 11th ed. (St. Louis: Mosby, 2002), 290–315.
2. Characklis WG, Marshall KG. *Biofilms.* (New York: Wiley, 1990), 3–15.
3. Shearer BG. Biofilm and the dental office. *JADA* 1996; 127:181–89.
4. Mayo JA, Oertling KM, Andrieu SC. Bacterial biofilm: A source of contamination in dental air-water syringes. *Clin Prev Dent* 1990; 12:13–20.
5. Williams JF, Johnston AM, Johnson B, Huntington MK, Mackenzie CD. Microbial contamination of dental unit waterlines: Prevalence, intensity and microbiological characteristics. *JADA* 1993; 124:59–65.
6. Pankhurst CL, Philpott-Howard JN. The microbiological quality of water in dental chair units. *J Hosp Infect* 1993; 23:167–74.
7. Williams HN, Kelley J, Folineo D, Williams GC, Hawley CL, Sibiski J. Assessing microbial contamination in clean water dental units and compliance with disinfection protocol. *JADA* 1994; 125:1205–11.
8. ADA Council on Scientific Affairs. Dental unit waterlines: Approaching the year 2000. *JADA* 1999; 130:1653–64.
9. Mills SE. The dental unit waterline controversy: Defusing the myths, defining the solutions. *JADA* 2000; 131:1427–41.
10. Christensen RP, Ploeger BJ, Hein DK. Dental unit waterlines: Is this one of dentistry's compelling problems? *Dent Today* 1998; 17:80–7.
11. Kohn WG, Collins AS, Cleveland JL, Harte JA, Eklund KJ, Malvitz DM, Centers for Disease Control and Prevention (CDC). Guidelines for infection control in dental health-care settings—2003. *MMWR Recomm Rep* 2003; 52(RR-17):1–61.
12. Karpay RI, Plamondon TJ, Mills SE, Dove SB. Combining periodic and continuous sodium hypochlorite treatment to control biofilms in dental unit water systems. *JADA* 1999; 130:957–65.
13. Van der Leeden F, Troise FL, Todd DK. *The water encyclopedia,* 2nd ed. (Chelsea, MI: Lewis Publishers, 1990), 446.
14. Miller CH. Microbes in dental unit water. *J Calif Dent Assoc* 1996; 24:47–52.

15. Williams JF, Molinari JA, Andrews N. Microbial contamination of dental unit waterlines: Origins and characteristics. *Compend Contin Educ Dent* 1996; 17:538–58.
16. Barbeau J, Tanguay R, Faucher E, et al. Multiparametric analysis of waterline contamination in dental units. *Appl Environ Microbiol* 1996; 62:3954–9.
17. Walker JT, Bradshaw DJ, Bennett AM, Fulford MR, Martin MV, Marsh PD. Microbial biofilm formation and contamination of dental-unit water systems in general dental practice. *Appl Environ Microbiol* 2000; 66:3363–67.

chapter seven

Biofilms in hospital water distribution systems

Judy H. Angelbeck, Kirsten M. Thompson, and Scott L. Burnett

Contents

Overview

Nosocomial infections, also referred to as hospital-acquired infections, account for more than ninety thousand deaths in the United States and cost the U.S. healthcare system $4.5 to 5.7 billion [1]. The causes of these infections are primarily attributed to breaches in hospital infection control practices, such as staff hand washing, ensuring the sterility of medical equipment, and providing clean environmental surfaces. Rarely is hospital water considered a source of exposure increasing the risk of nosocomial infection.

Anaissie et al. in a 2002 review identified hospital water as an unrecognized source of nosocomial infections, estimating that fourteen hundred deaths occur each year in U.S. hospitals as a result of waterborne nosocomial pneumonia caused by *Pseudomonas aeruginosa* alone [2]. This group identified the primary cause of diminished water quality as the buildup of biofilm and the corrosion in the distribution lines and tank surfaces of the hospital water distribution system [2].

While there may be many factors contributing to the poor recognition of hospital water, much less biofilm, in water distribution systems as a source of nosocomial pathogens, Costerton and Stewart [3] noted in their *Scientific American* review that scientists and healthcare professionals often envisioned bacteria as single cells that float through a watery environment. This could lead to an underappreciation of the more likely bacteria environment in colonies in extensive gooey biofilms coating surfaces.

A number of factors may contribute to the formation and maintenance of biofilm in hospital water distribution systems and tanks. These include the complex multistory and often aging distribution systems with corroded pipes and areas of water stagnation. Biofilm may become dislodged from pipe surfaces because of increased water demand during summer months, resulting in higher water flow rates and increased shearing forces. In addition, during periods of facility construction and remodeling, biofilm disturbance and release due to direct mechanical contact with the pipes can occur. The occasional use on water from less frequently accessed sites, combined with areas of stagnant water, can further add to opportunities for the release of biofilm into the water stream [2].

Patient exposure to waterborne microorganisms in the hospital occurs while showering, bathing, drinking water, ingesting ice, or inhaling water aerosols and droplets. It can also occur through contact with contaminated medical equipment such as tube feed bags, flexible endoscopes, and respiratory equipment that has been rinsed with tap water. The hands of healthcare personnel washed using tap water can also lead to patient exposure [2] (see Figure 7.1).

Microorganisms found in potable water and associated with biofilm include three categories: bacteria, mycobacterium, and fungi. In Table 7.1, microorganisms in these categories are identified and aligned with the types of infections caused and how they were identified, along with the number of reports correlating infections found in patients with microorganisms found in hospital water. In addition, Anaissie et al. identified several instances where antibiotic-resistant waterborne pathogens were associated with nosocomial infections [2].

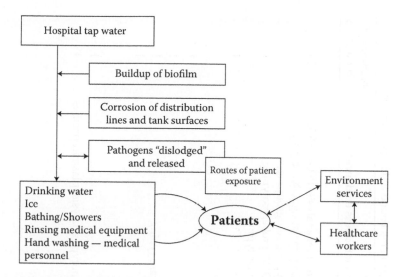

Figure 7.1 Schematic representation of the potential sites of waterborne microorganism sequestration within the plumbing in the hospital. (From Pall Corporation. With permission.)

Patients at high risk of infection due to waterborne microorganism include those who are immunocompromised as a consequence of their diseases or the treatment for their diseases. These include:

- AIDS patients
- Organ transplant patients
- Oncology patients
- Neonates
- Burn patients
- Critically ill patients in an ICU setting [4]

Table 7.1 identifies the number of reports associated with particular types of waterborne pathogens reported to have caused nosocomial infections in hospitalized patients. In addition, the table identifies the type of testing done to identify the microorganisms at the hospital water source and from the infected patient.

Controlling or eliminating biofilms in hospital water distribution systems is challenging, and systemic water treatments vary in efficacy and cost and may in some instances contribute to the problem by further corroding pipes. More recently, the use of point-of-use bacterial retentive filters has been reported in the literature to offer another tool to provide a new form of protection for patients, staff, and the hospital environment. Microorganisms in biofilms can be very resistant to treatments like chlorination.

Table 7.1 Evidence Correlating Nosocomial Infections with Microorganisms Found in Hospital Water

Organism	Site of infection	Molecular relatedness evidence[a]	Number of reports
Bacteria			
Pseudomonas aeruginosa	Blood, CVC, lungs, peritoneum, sinuses, trachea, urine	PCR, DNA macrorestriction analysis, PFGE, ERIC-PCR, RAPD, DNA fingerprinting, DNA typing, serotyping, phage typing, serogrouping, genotyping, ExoA DNA probe, biotyping, electrophoretic esterase typing	10
Stenotrophomonas maltophilia	Blood, peritoneum, respiratory tract, skin, stools, throat, trachea, urine	PFGE, RAPD	4
Serratia marcescens	Eye, stools	PFGE	1
Acinetobacter baumannii	Skin, wound	PFGE, biotyping	1
Aeromonas hydrophila	Blood	Electrophoretic esterase typing	1
Chryseobacterium spp.	Blood	AP-PCR	1
Mycobacterium spp.			
Mycobacterium avium	Disseminated	PFGE	1
Mycobacterium fortuitum	Disseminated, respiratory tract, sputum, sternal wound infection, wound	AP-PCR, PFGE, phenotype analysis, plasmid profiles	4

Mycobacterium xenopi	Various, spine	PCR-based techniques, chromosomal restriction fragment patterns	2
Mycobacterium kansasii	Abscess, blood, bone, sputum, stomach, urine	RFLP, PFGE	1
Mycobacterium chelonae	Sternal wound infection, prosthetic valve	Electrophoresis of enzymes, plasmid profiling	1
Fungi			
Fusarium solani	Disseminated	RFLP, RAPD, IR-PCR	1
Exophiala jeanselmaei	Disseminated	RAPD	1
Aspergillus fumigatus	Lungs	PCR, SSPD	1

Source: Adapted from Anaissie et al. [2].

[a] Abbreviations: AP = arbitrarily primed; PCR = polymerase chain reaction; PFGE = pulse field gel electrophoresis; ERIC-PCR = enterobacterial repetitive intergenic consensus sequence PCR; RAPD = random amplified polymorphic DNA; ExoA = exotoxin A; RFLP = restriction fragment length polymorphism; AFLP = amplified fragment length polymorphism; IR = interrepeat; SSPD = sequence-specific DNA primer analysis.

This chapter will provide more evidence demonstrating that biofilms in hospital water distribution systems are a source of risk, and a review of methods to control biofilms and potential exposure to biofilm microorganisms at the point of water use in the hospital.

Biofilm formation in hospital water distribution systems

Biofilm formation in hospital water distribution systems is similar to other water distribution systems; however, the potential for adverse effects of contaminated water for an immunosuppressed patient population is quite significant. While transient organisms within the water supply may vary, the organisms inhabiting the biofilm are expected to be more constant. In fact, the bacterial population in biofilm of water distribution systems appears to be similar throughout the United States. There have been no noted dominant populations due to unique water sources or treatment types [5].

Biofilms may appear as a patchy mass in some pipe section or as a uniform film along the inner walls of a storage tank. They may consist of a monolayer of cells in a microcolony or can be as thick as 10 to 40 mm. In the distribution system, attachment surfaces are typically pipe surfaces, sediments, tubercles, water storage tanks, valves, gaskets, and recesses. Biofilms are formed when pioneering organisms enter the distribution system through the municipal water supply and become entrapped in slow-flow areas or dead-end sections.

Growth is slow at first, but as the microcolony sites attract other microorganisms whose nutritional needs are met by the by-products of metabolism of the pioneers, the biofilm becomes quite diverse with a variety of bacteria, fungi, and protozoans. Species of *Pseudomonas, Flavobacterium, Arthrobacter, Acinetobacter, Sarcina, Micrococcus, Proteus, Bacillus, Actinomycites,* and some yeast frequently colonize biofilm in hospital water, and some have been acknowledged as opportunistic pathogens to immunocompromised patients.

Nutritional requirements

Trace concentrations of nutrients are a major factor in biofilm formation. Carbon, nitrogen, and phosphate are critical. An interesting factor to note is that oxidizing chemistries used for disinfection convert more of the nonbiodegradable organics to assimilable organic carbon; therefore, more nutrients may be available for the surviving organisms within the biofilm community after disinfection.[5]

Physical requirements

Structure

The physical structures of the water distribution system within the hospital include areas where stagnant water may accumulate, which encourages sediment and tubercle deposits along with microbial colonization. Sediment and pipe corrosion tubercles are attractive habitats, as their structure allows for microbial attachments and protection as well as nutrient entrapment [5].

Biofilm is generally considered to become a water quality problem in those water systems that have had pipes in service for over 50 years, but can easily become an issue in newer structures as well. Pipe condition is affected by not only composition, but also maintenance practices, as well as the chemistry of the water itself.

Structure and temperature

While very few studies dissect hospital biofilm specifically, models have been used to examine the role of biofilm in hospital water quality.

This study included the development of a model using filter-sterilized tap water as the sole source of nutrients for a mixed population of bacteria, fungi, and protozoa. Of specific interest to the investigators was the presence of *Legionella pneumophila* in both planktonic and biofilm phases. Copper tubing was compared to two plastics: polybutylene and chlorinated polyvinylchloride. Temperatures examined were 20, 40, 50, and 60°C [6].

The model included a simulated storage tank that was used to seed a constant inoculum into the simulated distribution system, which was used to generate biofilms. Coupons of the different plumbing materials were suspended within a separate model, and removed for observation at the same intervals.

L. pneumophila was present in the biofilms on the plastic materials as a proportion of the total biofilm flora, and very few, if any, *Legionella* were detected in any of the biofilms sampled on copper, even at the optimum growth temperature of 40°C. *L. pneumophila* did, however, survive in biofilm on the plastic materials at temperatures that are normally inhibitive of this organism in solution: 20 and 50°C. The investigation also demonstrated that copper tubing may be an ideal plumbing material, as the leaching of low levels of copper ions into the culture reduced the planktonic populations. In addition, the accumulation of these ions on other surfaces suggests that the use of this material may not only inhibit biofilm formation within the tubing, but also impart some antimicrobial resistance to other noncopper components.

Hydrophobicity

Cell surface hydrophobicity is a major parameter controlling the adhesion of bacteria to surfaces, and the affinity of bacterial lectins for carbohydrates is also important in biofilm formation, which may be observed by hemagglutination [7]. Hydrophobic interactions have been studied in environmental microbiology, while specific interactions such as the hemagglutination ability have been mostly explored in medical microbiology [8].

Water pressure

Key to a protected distribution of water supply in the hospital is maintaining an adequate positive water pressure (20 psi or more) throughout the entire pipe network. Water hydraulics can contribute to the release of biofilm organisms and their migration within the pipe network. Sudden restoration in water pressure causes shearing of biofilm, with organisms released into the water supply [6].

The primary cause of diminished water quality is the buildup of biofilm and the corrosion of distribution lines and tank surfaces, resulting from poor design or aging of distribution systems and water stagnation. Increased water demand during summer or construction activity increases flow through stagnant pipelines, dislodging organisms from biofilms and releasing them into the water supply [2].

Because water may remain stagnant for weeks or months in the pipes during construction of new hospitals, it is recommended to take specific measures (superheating, flush, cultures of the water) before the start of clinical activities [9].

Temperature

Unlike residential water tanks, which are based on flow-through systems, large buildings such as hospitals are required to have recirculating hot water systems, which are associated with a significant increase in risk of colonization and amplification of these organisms [2].

Conditions known to be favorable for the multiplication of *L. pneumophila* include stagnating warm water. Studies on both hot and cold water supplies have shown that *L. pneumophila* is relatively uncommon in cold water but has been isolated frequently from hot water systems. Continually keeping tanks on line by not allowing any downtime has been shown to keep *L. pneumophila* concentrations at low levels for prevention of infection in high-risk populations [10].

Most of the previous work on the optimum temperature range for *L. pneumophila* growth was performed on liquid cultures. From this information, it was determined that the organism would rapidly decline at temps above 50°C. Recommendations included keeping hot water stored above 50°C and cold water stored below 20°C. However, the organism has been shown to be persistent despite heat treatment, to include protection within biofilm on dead ends of pipe. The maintenance of the viability of *L. pneumophila* at 50°C suggests a protective factor from the biofilm at higher temperatures. This finding is significant as hot water systems that are not maintained and may fall a few degrees may be susceptible to *L. pneumophila* in biofilms [6].

Evidence of colonization in hospital water distribution systems

Colonization of hospital water distribution systems is well known and widespread. For this reason, some advocate the avoidance of tap water altogether for high-risk patients [2].

As the municipal water appears to be generally free of *Legionella*, the frequent contamination of hospital water systems remains perplexing. Hospital-acquired Legionnaires' disease does not correlate with the concentration of *Legionella* in water samples, but rather the presence of the organism [11,12].

The reason bacterial concentrations are not predictive is that *Legionella* is not concentrated in water but is harbored in the biofilm consisting of sediment and detritus at each distal water outlet [7,13].

In addition, the biofilm, and especially the protozoonotic relationship between *Legionella* species and certain free-living protozoa, may affect normal disinfection practice. Viable *Legionella* can be recovered from protozoan cysts, which can resist 50 ppm chlorine. Furthermore, such *Legionella* may demonstrate an enhanced resistance to chlorine [14].

The complex biological relationship between *Legionella* and protozoa, which may be temperature dependent, is important. It appears that *Legionella* gain entry to domestic water supplies by continual seeding from ruptured protozoa, as there is a clear association between their isolation in cold water distribution systems and the isolation of protozoa generally known to permit the intracellular growth of *Legionella*, including *Acanthamoeba, Echinamoeba, Hartmannella, Tetrahymena, Vahlkampfia,* and *Vanella* [14].

Water sampling alone is not predictive of the risk of *L. pneumophila* within a hospital water supply. When more sensitive methods for identifying Legionnaires' disease were employed, many more patients than in previous years were diagnosed with nosocomial *Legionella*. The hospital had experienced an outbreak of nosocomial Legionnaires' disease 12 to 14 years prior, and felt that in the interim years with less sensitive diagnostic tools, many Legionnaires' cases were undiscovered. Stored isolates from the previous outbreak investigations matched the clinical and environmental isolates from the current investigation, suggesting a continuous or recurrent colonization with the same organism over a 12-year period [15].

Fungi are important to consider as resident biofilm inhabitants. Results have demonstrated that potentially pathogenic molds inhabit the hospital water distribution system (the same type of molds inhabit the municipal water supply), and these molds become part of the hospital water system biofilm. Environmental sampling demonstrated that there appears to be a greater amount of airborne mold associated with areas of water usage, such as bathrooms, as opposed to areas such as hallways. Of the total number of water samples, 25% yielded *Aspergillus* species, which are common opportunistic pathogens [16].

Healthcare-associated infection linked to biofilm in hospital water

Beyond *Legionella* infection, gram-negative bacilli linked to waterborne infection include *Acinetobacter* species, *Burkholderia*, *Chryseobacterium*, *Ewingella*, *Pseudomonas*, *Serratia*, and *Stenotrophomonas*. *Mycobacterium* other than *M. tuberculosis* are also commonly found in water and may pose an infection risk to the immunocompromised. *Aspergillus* and *Fusarium* species have been isolated from hospital water distribution systems and pose a risk for healthcare-acquired infection [17].

Whirlpools

Commercial and residential whirlpools recirculate water that is treated with disinfectants; however, biofilm containing *P. aeruginosa* can proliferate quickly if the disinfectant used is below the recommended concentration [18].

During a 14-month period, immunocompromised patients were acquiring serious bloodstream and pneumonia infections caused by a single strain of multiply resistant *P. aeruginosa*, which was linked to the drain in a whirlpool bathtub on the unit. The water from the faucet was not contaminated, but became contaminated as the water came in contact with the drain as the bathtub was filled.

The drain closed approximately 2.54 cm below the surface of the tub. The area of the drain that was connected to the tub stayed moist, providing an excellent environment for the *Pseudomonas aeruginosa* to form colonies under a slime layer. When water was added to the tub, some of these colonies became dislodged and contaminated the bathwater—despite cleaning with quaternary ammonium compound, which was unable to penetrate the slime layer, and physical abrasion was not possible due to the drain cover.

Ice machines

Although ice machines are cold water outlets, several design features provide optimal conditions for microbial growth and formation of biofilm. The cold water passes through small-lumen, flexible tubing that is positioned close to the condenser/compressor. The heat warms the cold stagnant water to allow for bacterial growth and subsequent biofilm formation, which may be a source of many bacteria, to include *Legionella* [17].

Strategies to mitigate the risk of exposure

Control of biofilms in hospital potable water distribution systems

Several approaches to the prevention, control, and reduction of biofilm communities in hospital water distribution systems have been investigated. These include the use of heated water [19] and a number of chemical approaches, including free chlorine [20], monochloramine [21,22], chlorine dioxide [23,24], metal ions [25,26], and ozone [20,27] (see Table 7.2).

Table 7.2 Systemic Disinfection Approaches for Hospital Water Distribution Systems

Method	Efficacy/advantage	Disadvantage	References
Thermal treatment	Can reduce biofilm populations; inexpensive; requires no specialized treatment	Short-term benefit; risks associated with scalding	19, 20
Hyperchlorination	Can reduce biofilm populations; inexpensive	Recolonization can occur; limited efficacy; corrosivity to pipes	20, 29
Chlorine dioxide	Efficacious against biofilm populations	Recolonization can occur; safe handling precautions; corrosivity to pipes	23, 24
Monochloramine	Can reduce biofilm populations; less reactive to organic material than hypochlorite	Recolonization can occur; restricted availability to healthcare facilities	21, 22
Copper/silver ionization	Demonstrated efficacy against *Legionella* biofilm in laboratory; residual effectiveness	Relatively slow activity; pH dependent; reactive with chlorine	25, 26, 29
Ozonation	Can reduce biofilm populations; inexpensive; generally more efficacious than hypochlorite	Specialized equipment necessary	20, 28

Systemic control of biofilm formation and growth is difficult and often provides a short-term reduction in biofilm-associated microbial populations. Several factors are encountered in a water supply system that contribute to the limited success of systemic disinfection technologies. The ability of an adequate concentration of an antimicrobial chemistry or a lethal thermal dose (i.e., heated water) to reach all locations within the system is limited. The accumulation of scale, mineral deposits, and the presence of pits and crevices formed by corrosion may provide protection to microbial biofilms from physical removal or chemical inactivation, as well as a harborage site for continued proliferation. Also possible is the development and persistence of biofilm in dead ends and other locations of stagnation. In some cases, repair or replacement of the system has been shown to be effective but costly [2].

The Centers for Disease Control and Prevention (CDC) has published guidelines for the systemic treatment of hospital water distribution systems based on surveillance information targeting *Legionella* [28]. Chlorine is the most widely used antimicrobial chemistry for hot and cold water systems. Available chlorine levels should be maintained at 1–2 mg/L (1–2 ppm) at the tap [27]. Because free chlorine is highly reactive to organic material, consideration should be given to the degree to which a system is fouled with biofilm. In general, free chlorine at the concentration limits described is not efficacious in reducing microbial populations in a biofilm community. Water heated to 160 to 170°F (71 to 77°C) is another approach to disinfection of potable systems [28]. Maintaining temperature throughout the system can be difficult. In addition, the energy costs associated with this approach as well as the risk of accidental scalding injury of patients and staff present significant challenges in the practical implementation of thermal treatment [2].

Monochloramines have received much interest in recent years for use in the systemic treatment of hospital water systems. Unlike the hypochlorites, monochloramine is less reactive to organic material and is not associated with the formation of disinfection by-products such as trihalomethanes and haloacetic acid. In a 2-year prospective study of water and biofilm samples from the water supply sytems of fifty-three buildings, prevalence of *Legionella* declined from 60% before monochloramine treatment to 4% afterward [22].

Point-of-use interventions to minimize patient exposure

Treatment of hospital drinking water at the point of use can be an effective approach to mitigate the risk of nosocomial infection, especially when used in conjunction with a systemic disinfection treatment. Physical approaches to the removal or inactivation of suspended microorganisms, including boiling, ultraviolet light exposure, and filtration, have been implemented. Point-of-use interventions target microorganisms that have been dispersed from biofilm communities within the water distribution system by shedding. Such shedding events are commonly encountered in the life cycle of a biofilm and can be exacerbated by the application of systemic treatment, like the dosage of the system with biocides. Point-of-use treatment of hospital drinking water should be used in adjunct to systemic treatments.

Providing patients with cooled, boiled water has been recommended in some transplant programs [29]. While this approach is effective in eliminating even the most resistant microorganisms of concern, including protozoan parasites, there are several disadvantages to consider. Boiling is time- and energy-consumptive, and caution must be practiced to avoid the risk of scalding or burning patients and staff. Diminished organoleptic quality of boiled water has also been reported [2].

The exposure of water to an in-line source of ultraviolet (UV) light (254 μm) has also been explored [20] with favorable results. Hall et al. [30] reported the absence of water samples positive for the presence of *Legionella* (*n* = 930) analyzed in a hospital incorporating a UV light apparatus in its potable water distribution system. The hospital reported the lack of a clearly documented case of legionellosis for 13 years. Turbidity and large clumps of biofilm and other debris in the water stream may shield microorganisms from UV light, thus limiting its efficacy.

The incorporation of point-of-use filtration is an effective approach in ensuring the safety of water in hospitals. Water is passed through a 0.2 μm filter, which is housed in a disposable cartridge. The cartridges are located at the terminal of the plumbing system (e.g., shower heads, faucet taps, ice machines). Typically they need to be changed out every 7 days, depending on the manufacturer. Several studies and cohort investigations have evaluated the efficacy of filtration in hospital environments. Sheffer et al. [31] evaluated water filtration as an intervention approach within a hospital with chronic *Legionella* contamination. The authors concluded that point-of-use filtration cartridges could prevent exposure for high-risk patients to waterborne pathogens. Filtration complements systemic water disinfection technologies by trapping planktonic microorganisms that have shed from biofilm communities or otherwise evaded disinfection treatment. The implementation of point-of-use filters has been reported to decrease the incidence of *Legionella* infection from 23% to 15% after installation [31]. The incidence of waterborne *Pseudomonas aeruginosa* before and after a hospital-wide implementation of point-of-use filters fell from 2.5 to 0.9 infections per month [33]. Point-of-use filters may serve as a successful adjunct to routine systemic water disinfection treatment regimens for the reduction of exposure to waterborne pathogens. Care must be exhibited to routinely replace filter cartridges as recommended by the manufacturers.

Judy H. Angelbeck, PhD, is senior vice president of Pall Life Sciences. She has both North American and global medical product experience from more that twenty-two years in marketing, regulatory, and general management. Dr. Angelbeck is currently working on development of new products for the somatic cell therapy market.

Kirsten M. Thompson is a technical support expert for Ecolab's healthcare division. She leads the technical service department, providing information pertaining to infection control and epidemiology in both acute and alternate healthcare markets.

Scott L. Burnett, PhD, is a scientist for Ecolab's food and beverage division, where he leads a research team with an aim of developing innovative antimicrobials for food surface and food processing environments.

References

1. Weinstein, R.A., Nosocomial infection update, *Emerg. Infect. Dis.*, 4, 416, 1998.
2. Anaissie, E.J., Penzak, S.R., and Dignani, C., The hospital water supply as a source of nosocomial infections; a plea for action, *Arch. Intern. Med.*, 162, 1483, 2002.
3. Costerton, J.W., and Stewart, P.S., Battling biofilms, *Sci. Am.*, 285, 74, 2001.
4. Ortolano, G.A., et al., Hospital water point-of-use filtration: A complementary strategy to reduce the risk of nosocomial infection, *Filtration Suppl.*, 1, 3, 2004.
5. Geldreich, E.E., Biofilms in water distribution system, in *Microbial quality of water supply in distribution systems*, ed. E.E. Geldreich, CRC Press, Boca Raton, FL, 1996, chap. 4.
6. Rogers, J., et al., Influence of temperature and plumbing material selection on biofilm formation and growth of *Legionella pneumophila* in a model potable water system containing complex microbial flora, *Appl. Environ. Microbiol.*, 60, 1585, 1994.
7. Ta, A.C., et al., Comparison of culture methods for monitoring *Legionella* species in hospital potable water systems and recommendations for standardization of such methods, *J. Clin. Microbiol.*, 33, 2118, 1995.
8. Fiorina, J.C., Weber, M., and Block, J.C., Occurrence of lectins and hydrophobicity of bacteria obtained from biofilm of hospital catheters and water pipes, *J. Appl. Microbiol.*, 89, 494, 2000.
9. Merlani, G.M., and Francoli, P., Established and emerging waterborne nosocomial infections, *Curr. Opin. Infect. Dis.*, 16, 343, 2003.
10. Ciesielski, C.A., Blaser, M.J., and Wang, W.-L.L., Role of stagnation and obstruction of water flow in isolation of *Legionella pneumophila* from hospital plumbing, *Appl. Environ. Microbiol.*, 48, 984, 1984.
11. Kool, J.L., et al., Hospital characteristics associated with colonization of water systems by *Legionella* and risk of nosocomial Legionnaire's disease: A cohort study of 15 hospitals, *Infect. Control Hosp. Epidemiol.*, 20, 798, 1999.

12. Best, M., et al., Legionellae in the hospital water supply. Epidemiological link with disease and evaluation of a method of control of nosocomial Legionnaire's disease and Pittsburgh pneumonia, *Lancet*, 2, 307, 1983.
13. Wright, J.B., et al., *Legionella pneumophila* grows adherent to surfaces in vitro and in situ. *Infect. Control Hosp. Epidemiol.*, 10, 408, 1989.
14. Patterson W.J., et al., Colonization of transplant unit water supplies with *Legionella* and protozoa: Precautions required to reduce the risk of legionellosis, *J. Hosp. Infect.*, 37, 7, 1997.
15. Lepinel, L.A., et al., A recurrent outbreak of nosocomial Legionnaires' disease detected by urinary antigen testing: Evidence for long-term colonization of a hospital plumbing system, *Infect. Control Hosp. Epidemiol.*, 19, 905, 1998.
16. Anaissie, E.J., et al., Pathogenic molds (including *Aspergillus* species) in hospital water distribution systems: A 3-year prospective study and clinical implications for patients with hematological malignancies, *Blood*, 101, 2542, 2003.
17. Squier, C., Yu, V.L., and Stout, J., Waterborne nosocomial infections, *Curr. Infect. Dis. Rep.*, 2, 490, 2000.
18. Berrouane, Y.F., et al., Outbreak of severe *Pseudomonas aeruginosa* infections caused by a contaminated drain in a whirlpool bathtub, *Clin. Infect. Dis.*, 31, 1331, 2000.
19. Ezzeddine, H., et al., *Legionella* spp. in a hospital hot water system: Effect of control measures, *J. Hosp. Infect.*, 13, 121, 1989.
20. Muraca, P., Stout, J.E., and Yu, V.L., Comparative assessment of chlorine, heat, ozone, and UV light for killing *Legionella pneumophila* within a model plumbing system, *Appl. Environ. Microbiol.*, 53, 447, 1987.
21. Kool, J.L., Carpenter, J.C., and Fields, B.S., Effect of monochloramine disinfection of municipal drinking water on risk of nosocomial Legionnaires' disease, *Lancet*, 353, 272, 1999.
22. Flannery, B., et al., Reducing *Legionella* colonization of water systems with monochloramine, *Emerg. Infect. Dis.*, 12, 588, 2006.
23. Walker, J.T., et al., Control of *Legionella pneumophila* in a hospital water system by chlorine dioxide, *J. Ind. Microbiol.*, 15, 384, 1995.
24. Srinivasan, A., et al., A 17-month evaluation of a chlorine dioxide water treatment system to control *Legionella* species in a hospital water supply, *Infect. Control Hosp. Epidemiol.*, 24, 560, 2003.
25. Goetz, A., and Yu, V.L., Copper-silver ionization: Cautious optimism for *Legionella* disinfection and implications for environmental culturing, *Am. J. Infect. Control*, 25, 449, 1997.
26. Rohr, U., et al., Four years of experience with silver-copper ionization for control of *Legionella* in a German university hospital hot water plumbing system, *Clin. Infect. Dis.*, 29, 1507, 1999.
27. Landeen, L.K., Yahya, M.T., and Gerba, C.P., Efficacy of copper and silver ions and reduced levels of free chlorine in inactivation of *Legionella pneumophila*, *Appl. Environ. Microbiol.*, 55, 3045, 1989.
28. Edelstein, P.H., et al., Efficacy of ozone in eradication of *Legionella pneumophila* from hospital plumbing fixtures, *Appl. Environ. Microbiol.*, 44, 1330, 1982.
29. Sehulster, L., and Chinn, R.Y.W., Guidelines for environmental infection control in health-care facilities, *Morb. Mort. Weekly Rep.*, 52(RR10), 1, 2003.

30. Hall, K.K., et al., Ultraviolet light disinfection of hospital water for preventing nosocomial *Legionella* infection: A 13-year follow-up, *Infect. Control Hosp. Epidemiol.*, 24, 580, 2003.
31. Sheffer, P., et al., Efficacy of new point-of-use water filters to prevent exposure to *Legionella* and waterborne bacteria, paper presented at Association for Practitioners or Infection Control, Phoenix, AZ, 2004, abstract 21793.
32. Hummel, M., Kurzuk, M., and Hetzer, R., Prophylactic and pre-emptive strategies for control of Legionnaire's prevention, poster abstract presented at the Fourteenth Annual Scientific Meeting of the Society for Healthcare Epidemiology of America (SHEA), Philadelphia, 2004, abstract 191.
33. Trautmann, M., et al., Tap water colonization with *Pseudomonas aeruginosa* in a surgical intensive care unit (ICU) and relation to *Pseudomonas* infection of ICU patients, *Infect. Control Hosp. Epidemiol.*, 22, 49, 2001.

chapter eight

Disinfection and its influence on the ecology of biofilms in drinking water distribution systems

Fernando M. G. Matias, V. Susan Springthorpe,
Syed A. Sattar, Marsha Pryor, and Graham Gagnon

Contents

Introduction

Any nonsterile wet or damp surface will grow a biofilm of microorganisms indigenous to that milieu. Therefore, all drinking water distribution system surfaces can be assumed to be coated to some extent by biofilms; this includes consumer taps. Biofilms often appear slimy on rocks, ceramics, etc., but in drinking water distribution systems with significant mineral content, the biofilm can appear as encrustation on the pipe surfaces, or may not even be visually apparent unless examined microscopically. There have been many arguments over what constitutes a biofilm, especially regarding the density and coverage (quantity) of microorganisms present. However, for most purposes it is simplest to think of a biofilm as any population of sessile microorganisms that grow attached to surfaces, and that can periodically be dispersed through any fluids passing over the surfaces, regardless of whether a film is apparent.

Biofilms are now believed to be the natural growth habit of bacteria, with the free planktonic cells, with which we are so familiar in laboratory cultures, likely to be merely a dispersal phase during bacterial growth and development (O'Toole et al., 2000). In addition to cells, biofilms include an extracellular matrix that gives them structure. Although this extracellular matrix was originally believed to be mainly polysaccharide with multiple functional roles (Wotton, 2004), it is now known to contain considerable quantities of nucleic acids and proteins as well (Lawrence et al., 2003; Böckelmann et al., 2006). Biofilms are usually discussed in the context of solid surfaces; in reality, they occur at all interfaces, including those between fluids and air (Wotton and Preston, 2005), as well as on suspended particulates. Most of what we know of biofilm formation, architecture, dispersal, and other characteristics has been derived from laboratory studies of simple monoculture biofilms on relatively uniform surfaces. Natural biofilms are considerably more complex and may not conform entirely to the textbook stereotype.

Biofilm as a microbial ecosystem

Drinking water is relatively low in nutrients, and favors the growth of a range of microorganisms adapted to low-nutrient environments. However, as growth occurs, the bodies of the colonizers and their wastes become nutrient sources concentrated at the biofilms, and these permit the growth of more colonizers and eventually a wider range of microorganisms. As the biofilm organisms become numerous, they attract free-living protozoan predators that graze the biofilm surface and control its depth (Zubkov and Sleigh, 1999; Kadouri and O'Toole, 2005; Huws et al., 2005; Salcher et al., 2006). The supposition is that the predators choose mainly to feed on cells, but it is also clear that the extracellular material can be a prime or, in some cases, preferred nutrient source (Joubert et al., 2006). Biofilm bacteria can also resist protozoan grazing by a variety of adaptation mechanisms that operate pre- or postingestion (Matz and Kjelleberg, 2005). Of most concern in this context are the postingestion adaptations that allow human pathogens to persist inside protozoa, including in encysted forms, and thereby potentially escaping disinfection (King et al., 1988; Snelling et al., 2005b). From the standpoint of drinking water, the most prominent and well-recognized example of this is *Legionella pneumophila* persistence in a variety of protozoa (Thomas et al., 2004; Garcia et al., 2007). There is also evidence that bacteria that become endosymbionts of free-living protozoa can potentially evolve to greater virulence along with their host (Albert-Weissenberger et al., 2007; Hilbi et al., 2007). However, other important pathogens are known to have similar mechanisms to escape predation, and it is likely that biofilms will play important roles, for example, in the persistence of *Campylobacter* (Snelling et al., 2005a) and *Helicobacter* (Azevedo et al., 2003, 2006) in drinking water environments.

Thus, the drinking water distribution system (DWDS) biofilm is really a specialized microbial ecosystem adapted to conditions present in the DWDS of each individual water utility. DWDS biofilm contain many bacteria, including actinomycetes, some fungi, often algae, and a variety of bacteriovores, including motile eukaryotic invertebrates (Thomas et al., 2004) and lytic bacteria (Kadouri and O'Toole, 2005). Microbial viruses are likely to be active in such environments, and it is also possible for human pathogenic viruses to become trapped in biofilms (Skraber et al., 2005), but numbers of the latter will usually be very low; however, vulnerable water sources that do not receive full conventional treatment could be a factor in virus transmission.

Depending on the water composition, mature DWDS biofilms may become dominated by relatively few species, with other species present playing a relatively minor numerical role unless disturbance makes the biofilm conditions more favorable for their multiplication. An example of such a situation may occur if one bacterium cannot compete for a particular nutrient with a dominant species, but a change of treatment regime or pipe material results in that dominant bacterium decreasing and another dominant one arising. Biofilm inhabitants are there because they play a role in the complex community that is present. Although it is beyond the scope of this chapter to discuss fully the properties of biofilms and communication among biofilm inhabitants, a number of good reviews are available that explore these issues in depth (Shapiro, 1998; Davey and O'Toole, 2000; Stoodley et al., 2002; Zhang and Dong, 2004; Parsek and Greenberg, 2005; Shiner et al., 2005; Reading and Sperandio, 2006).

There are several advantages to microorganisms living in a biofilm. These include nutrient acquisition, formation of mixed-species assemblages with the combined abilities to degrade complex and potentially recalcitrant substrates (Szewzyk et al., 2000), and the ability of close contact with similar neighbors to coordinate their activities to greater effect (Juhas et al., 2005). In addition, the biofilm growth habit can protect from toxins that can bind to the extracellular matrix, or that are detoxified by neighbors. In addition to disinfectant residuals and their by-products in the DWDS, numerous toxins in source waters can be bioaccumulated from the flowing water by the stationary biofilm. Biofilms growing in DWDS are essentially specialized, and their persistence is encouraged by the long pipes on which they grow, and a continuous source of flowing water that brings nutrients and disperses waste products. These relatively ideal conditions permit their continuous culture and the selection of those microorganisms that are best suited to the prevailing conditions.

What controls biofilm composition?

To the extent that prevailing conditions in the DWDS vary among utilities, one can expect that the biofilms will themselves be of different composition (Norton and LeChevallier, 2000), although it is rare for more than one source water or utility to be studied. In spite of such differences, the main groupings of microorganisms that carry out basic functions of nutrient cycling will be expected to be repeated, even if the microorganisms that comprise them vary somewhat. When undisturbed for prolonged periods, with a continuous and stable water supply, the biofilm can become quasi-stable over time. However, biofilms in natural water sources and drinking water derived from them are subject to the hydrogeological, climatic, anthropogenic, and seasonal influences that control the quality of the water and give it its physical, chemical, and biological characteristics. Such variations can be expected to be reflected by changes in biofilm composition. Most emphasis is placed on the bacterial composition of the biofilm community due to the possible presence and public health significance of frank or opportunistic bacterial pathogens. There is nevertheless a need to understand the types of microorganisms present and their functional roles. This includes not only bacteria but also how environmental variables alter the composition of other microbial classes, and the presence or absence of the various types of predators.

Many factors control biofilm formation and composition. First, microorganisms must be seeded into the DWDS, most likely from the source water after escape of the multibarrier treatment process. Passage through filters during filter ripening may be one of the common escape routes, with subsequent protection from the applied disinfectant due to intrinsic bacterial properties, or presence in a protected environment, such as particulates or inside protozoa. If bacterial colonizers escaped disinfection initially, and then came to proliferate in the presence of a disinfectant residual, they inevitably have a certain degree of resistance to the concentrations of applied chemicals present at the biofilms. Once biofilm organisms are seeded into the proximal DWDS, they can then spread to distal ends of the system. A summary of the potential influences on biofilm formation and composition is shown in Table 8.1. Biofilm composition can also be influenced by downstream incursions from pressure transients (LeChevallier et al., 2003), pipe breaks, and deteriorating infrastructure that may admit to the DWDS microorganisms from soil, storm water, or raw sewage.

Table 8.1 Influences and Effects on Biofilm Composition

Influence	Likely effect
Source water composition	Controls microorganisms available for biofilm formation
Treatment regime— coagulation and settling	Efficiency controls proportion of source water bacteria and particulates presented for filtration
Treatment—filtration	Efficiency, especially during ripening, controls proportion of bacteria and particulates presented to disinfectant
Treatment—disinfectant	Selects types and numbers of viable bacteria entering or remaining in DWDS biofilms
Bacterial properties	Controls ability to attach to DWDS surfaces
Nature of pipe materials	Selects types of microorganisms that prosper in the biofilm
Availability of nutrients and minerals from plant effluent	Helps to maintain the dominant species within the biofilm; changes in water quality can change biofilm composition
Other species within the biofilm	Provide antagonistic competition, protection, cooperation, or nutrition, predation; species occupy niches that allow them to remain as a part of the biofilm; functions performed by each help to control or encourage the other bacteria present

Do all the bacteria admitted to the DWDS necessarily end up in biofilms? On a numerical basis, this is certainly not true; but even if one considers whether all available types are present in biofilms, then there is no evidence or consensus for this. The mix of viable microorganisms present in DWDS is controlled by the physical, chemical, and biological composition of the finished water and by the treatment regime as well as the hydraulic conditions and the integrity of the distribution infrastructure. Upon entering the DWDS, the fraction of those microorganisms that initiate or participate in biofilm formation will again be controlled by the water composition and hydraulic conditions, but now the nature of pipe material, and the properties of the bacteria that allow them to attach to it, play an important role (Norton and LeChevallier, 2000; Batté et al., 2003; Kalmbach et al., 1997; Matias et al., 2006). It is also important to note that nonviable organisms and microbial waste products may become trapped in the biofilms, and these are potential nutrient sources for the growing biofilm.

Biofilms are dynamic structures that change temporally and are also diverse spatially. The complexities offered by many different collaborating and competing species in natural DWDS biofilms are not readily understood by experimental studies, nor can laboratory studies capture the changes that occur due to seasonal or temporal changes in water quality. Thus, studies of DWDS biofilms should ideally be performed in as realistic a scenario as possible, with biofilm samples collected from a full-scale plant, a slipstream of full-scale plant effluent, or perhaps a pilot system using identical sources and treatment systems. Superimposed on the inherent variability are successional mechanisms operating in any functioning ecosystem, and longitudinal changes as water ages in the DWDS and the disinfectant residual declines. Localized conditions of corrosion at the pipe walls, incursions from outside the DWDS, or even migration of motile organisms at the pipe walls from distal sites may cause additional local disturbance in the biofilm. More extensive disturbance or dislodging of biofilm may arise from hydraulic alterations, such as a sudden increase in velocity from flushing the main lines or an oscillating pressure ("water hammer") caused by changing valve positions.

It remains expensive to conduct detailed biofilm studies, but with rapid advances in environmental microbiology techniques, it may be possible in the future for many utilities to acquire a good understanding of their DWDS biofilms. Among the most interesting questions about biofilm in drinking water is: How can disinfection regimes affect biofilm composition? We have asked this question previously in our research, and the purpose of this chapter is to bring together information from our own work and other published studies to provide a synthesis of current knowledge in this area. This research is in the very early stages, and so the information available constitutes just glimpses into the complexities of drinking water biofilms. It is necessary to say at the outset that the information is focused almost exclusively on the bacterial composition. Our knowledge of whether disinfectants can affect the eukaryotic species available is very limited, but the presence of large numbers of protozoa in locations such as cooling towers suggests that many of them may resist the levels of disinfection present in drinking water or applied in the cooling towers.

How can we understand the complex microbial community of biofilms? What techniques are available?

With current technologies, comprehensive biofilm studies are difficult to perform *in situ* and impossible to replicate exactly *in vitro*. However, there are a number of initiatives to try and gain insights into DWDS biofilms in general, or to specialized biofilm-based phenomena in particular, such as the nitrification that can occur within the DWDS, especially in chloraminated systems.

Theoretically, both traditional culture methods and molecular methods can be used to explore microbial diversity, but each has its own biases and limitations and provides only partial information. During culture, the bias is created by the growth substrates and conditions provided in the culture media, which are usually suitable for only a small minority of the viable population, including media designed to enumerate heterotrophs (Allen et al., 2004; Bartram et al., 2004; Farleitner et al., 2004). Even during comprehensive culturing approaches with multiple media we see perhaps, at maximum, 1% of the microbial biomass. The cultivable fractions of cells in complex communities, such as biofilms, are not representative of either the abundance or the diversity present (Emtiazi et al., 2004; Kent and Triplett, 2002; McBain et al., 2003; Penna et al., 2002; Vincke et al., 2001), and many cells may be viable but refractive to culture. Failure to culture the full range of bacteria present is due to our limitations in understanding of how to produce the necessary growth conditions (Kent and Triplett, 2002), the interdependency of different organisms upon each other (Amann et al., 1995), and that environmental stress situations that may induce physiological and morphological changes in many bacteria (Emtiazi et al., 2004). Moreover, even if we knew how, it would be impractical to use enough different culture media to grow and identify all the community members. Nevertheless, the advantages of culture allow particular organisms to be sought relatively simply, obtained in pure culture, characterized by morphologic, biochemical, and physiological properties, and to be used for subsequent studies.

The limitations of culture have, however, spawned a variety of culture-independent molecular methods that are better suited for characterization of complex microbial communities, but have the drawbacks that you cannot obtain viable organisms for further study and biochemical characterization. Perhaps one could then argue why not use both types of techniques in parallel. Certainly that can increase the information obtained, but it also greatly increases labor, and because it is unlikely that the populations will be directly comparable, it will still give only a partial community characterization.

Molecular analysis most often uses polymerase chain reactions (PCRs) to amplify phylogenetic markers from the extracted DNA of the microbial community; for bacteria, variable regions of the 16S rDNA gene are most frequently used. Characteristic DNA is present in each bacterial cell regardless of its physiological state, but in a complex environment such DNA must be isolated, separated, and identified (Kent and Triplett, 2002) by a variety of techniques (described below) in order to understand the composition of the bacterial population. Sequencing is used to identify microorganisms from the amplified phylogenetic markers (Emtiazi et al., 2004). This can be a disadvantage since many unknown bacteria can be encountered in complex communities, which may not find a match within the sequencing library. Also, characterization of unknowns would require studying their metabolic, morphologic, and physiological traits, which would require culturing the organism, even though this may not currently be possible. Short sequences (~200 bp) are usually obtained with this method, which also may not be specific enough to identify organisms down to the species level. Therefore, although successful, using a molecular biology approach only focuses on the identification of microbial diversity, and does not give any information on the complex dynamics of microbial interactions, or identifying traits of unknown organisms. Moreover, for molecular methods, the obvious bias is that the more numerically dominant genomes extracted are likely to be the most easily detected (Holben and Harris, 1995; Vincke et al., 2001; Wang and Wang, 1997), and microbes that form only a small proportion of those present may be missed. Further, less obvious and uncharacterized biases arise from differential levels of DNA extraction or preferential PCRs that could skew interpretation of the community structure. Culture-independent methods have nevertheless become the norm in characterizing the microbiology of complex environments (Brooks et al., 2003; Emtiaza et al., 2004; Kent and Triplett, 2002; Vincke et al., 2001), and are increasingly being coupled with other techniques to broaden the scope of the information they can provide. In the following sections we describe and critically discuss the types of methods that have been or can be used to explore microbial community structure. It is important to note that very few such studies have so far dealt with DWDS biofilms, and these also are discussed briefly below.

Polymerase chain reaction

PCR has been used to assess the microbial diversity of many environmental samples (Vincke et al., 2001; McBain et al., 2003), including drinking water (Emtiazi et al., 2004; Farleitner et al., 2004). This technique relies on the assumption that most sequences are complementary to the universal primers used in their amplification. However, it has been observed that as the database of 16S rRNA gene sequences has grown, new taxonomic groups have been discovered that are not amplified by these primers (Baker et al., 2003). Primers designed to be complementary to conserved regions of the groups present at the time of their production are not necessarily complementary to all the groups that are known today. However, if the main interest is with pathogens, opportunistic pathogens, and known bacteria, then most universal primers would have been created with these in mind.

Primers that include multiple nucleotides or inosine residues (Watanabe et al., 2001) at degenerate positions have been used to provide universal specificity. However, it has been found that this can lead to amplification of nontarget genes or domains (Baker et al., 2003). Inosine residues are also considerably more expensive than standard bases. Although more research can provide better universally conserved primers, it would be virtually impossible to obtain a set of perfect primers for the amplification of all prokaryotic 16S rRNA genes. Another means of avoiding the amplification of non-rDNA fragments, or fragments with improper sizes, is by using a "touchdown" PCR technique. However, this can be quite strenuous on primer binding and can lead to biases in amplifying only species with perfect primer matching. It has therefore been recommended by Watanabe et al. (2001) that the touchdown method only be used when single annealing temperature PCR produces unwanted amplification products.

Environmental samples require special consideration for DNA extraction since PCR inhibitors can be extracted along with the DNA (Vincke et al., 2001; Kent and Triplett, 2002; Braid et al., 2003). Water samples that may have low bacterial density are hard to extract and amplify, possibly requiring prior filtration of many liters to get a good representation of the population. However, PCR inhibitors can also be co-concentrated (Kent and Triplett, 2002; Ashbolt, 2003) and may need to be removed. There is also the issue of ensuring lysis of structurally different cells; this may require inclusion of a nonspecific mechanical lysis method, such as bead beating (Kent and Triplett, 2002). While this may help ensure that the diverse population is adequately represented, some characteristics of particular organisms or their nucleic acids may result in biases occurring during DNA extraction or PCR amplification, as mentioned above.

One of the greatest advantages of PCR is that the target organisms do not need to be cultured. This can lead to the detection of new uncultivable organisms (Ashbolt, 2003). It is also a very good tool for identifying pathogens within high numbers of background bacteria, such as in a biofilm. Using specific primers for known sequences, single species can be picked out of a biofilm that includes a wide variety of other bacteria (Ashbolt, 2003). Such known sequences can also be used in calibrated and quantitative PCR (Q-PCR) reactions to provide a quantitative measurement of how many copies of the template are in the sample. This is normally done in real time, and can provide an estimate of dominance in a sample.

Denaturing gradient gel electrophoresis and temperature gradient gel electrophoresis

One of the methods commonly used for analyzing community structure in conjunction with PCR is denaturing gradient gel electrophoresis (DGGE), or PCR-DGGE. In this procedure, after PCR amplification the DNA fragment obtained from all bacteria is of a single size based on the primers used. This DNA, however, has different sequences, depending on the species of bacteria, and it therefore needs to be separated using the sequence structure as the determinant. This is achieved using a linear gradient of DNA denaturants (urea and formamide) that influences the melting behavior of the amplicons when run on the gel, and therefore molecules with different sequences will stop migrating at different positions in the gel (Muyzer et al., 1993, 1998; Wang and Wang, 1997;

Muyzer and Smalla, 1998; Muyzer, 1999; Vincke et al., 2001; Watanabe et al., 2001; Werker and Hall, 2001; Ashbolt, 2003).

PCR-DGGE analysis of complex communities therefore produces bacterial community profiles, or fingerprints, which generally represent the most dominant or readily amplified members of the bacterial population (Gelsomino et al., 1999; Muyzer et al., 1993; Holben and Harris, 1995; Wang and Wang, 1997; Muyzer and Smalla, 1998; Vincke et al., 2001). The obtained bacterial profile can then be fairly easily compared to other bacterial profiles, which would allow easy identification of correlating or dissimilar organisms. This technique has been used for many sample types, including soils (Gelsomino et al., 1999; Kent and Triplett, 2002; Gonzalez et al., 2003), drinking water (Boe-Hansen et al., 2003; Emtiazi et al., 2004; Farleitner et al., 2004; Matias et al., 2006), and biofilms (Vincke et al., 2001; Boe-Hansen et al., 2003; McBain et al., 2003; Matias et al., 2006), among many others.

There are some problems that may restrict the interpretation of PCR-DGGE. As stated above, PCRs are rarely perfect, which means that some of the bands on a DGGE may not represent true bacterial amplicons but rather chimeric forms (Wang and Wang, 1997) or spurious by-products. There is also the issue that certain targets may be amplified more readily than others, which would distort the distribution of band intensities, and therefore dominance of organisms within the population cannot be correlated to the intensity of the banding pattern on the gel. This, however, has been shown to be a minor practical problem (Gelsomino et al., 1999; Kent and Triplett, 2002). It has also been shown that single bacterial types can produce more than one band on a DGGE profile, due to the presence of several slightly different copies of the 16S rRNA gene. There is also the possibility that multiple sequences may have a very similar melting behavior and show up as a single band, even though they are from different species (Gelsomino et al., 1999; Kent and Triplett, 2002), which would make direct sequencing from an excised band impossible. Most importantly, due to the competitive nature of the multitarget PCR, only a range of the most dominant bacterial types are detectable. Gelsomino et al. (1999) has shown evidence that bacterial populations that make up as little as 0.1% of the total community can be detected by PCR-DGGE, while Emtiazi et al. (2004) and Muyzer et al. (1993) report this figure to be 1% at the lowest. This is important in interpreting differences between profiles since any change in the abundance of a minor bacterial component in the community can impact the profile, while dominant members are not as affected. However, even with all these problems, PCR-DGGE has been shown to be a reproducible and robust strategy for the assessment of diversity in complex communities (Muyzer et al., 1993; Gelsomino et al., 1999). Identification of obtained bands can then be done by excision, reamplification, cloning, and sequencing, making essentially every band identifiable. However, these additional steps can be time-consuming, costly, and at times difficult.

Gonzalez et al. (2003) have described some modifications that may offer some advantages for the speed and cost of performing community analysis. Following DNA isolation, they have suggested that in addition to the regular DGGE universal primers, one should also separately PCR amplify using the universal primer pair 27F and 1497R. This provides a near-full-length 16S rRNA sequence from the DNA extract, which can then be cloned creating a 16S rDNA library. The library can then be cross-checked against the original ~200 bp amplicons on DGGE by performing nested PCR using the same DGGE primers on the clones. They also suggested performing the nested PCR on ten clones per reaction, in order to minimize lane usage on the DGGE. If matching bands are obtained, the clones would then be individually PCR amplified, and run against the original sample to identify which clone provided which band. That specific plasmid can then be sequenced, giving approximately full-length instead of the short ~200 bp sequences, and thus a better chance at unambiguous identification to the species level. Also, DGGE bands do not need to be excised, reducing contamination possibilities and removing a difficult and time-consuming set of steps. Furthermore, this method should also be considerably cheaper since cloning is done on the entire sample, and not on each individual band. In order to obtain good coverage of the overall population, this cloning step may be done several times; however, it is still much less than is required when individual bands are processed. The randomness involved in picking the correct clone with the excised band position is not required with this method, but there is still some chance involved in picking clones from the library and comparing them to the master profile. By adding ten clones per reaction, you can cut down on DGGE gel lanes being used, but for the first few batches of samples, they will need to be done separately as well, since all bands will be new and useful. Only after the consistently dominant bands are identified will the mixing of ten clones become useful in reducing the number of samples.

Gonzalez et al. (2005) has further offered a means for amplifying low copy number sequences that may be present in environmental samples. This consists of a whole genome preamplification with subsequent conventional PCR targeting specific genes. The sensitivity appears to be about tenfold higher but could certainly be useful when targeting genes specific to particular groups of microorganisms that may be present in low abundance.

Temperature gradient gel electrophoresis (TGGE) is very similar to DGGE, except that a temperature gradient is used to affect the mobility of the DNA traveling through the gel instead of denaturants. This technique is often more complicated than DGGE mainly because it is difficult to obtain a precise and stable temperature gradient. In DGGE the gradient is held solidly in place in the gel matrix, while in TGGE heating rods, or pumps, are used to provide a temperature gradient surrounding the gel. Although new, more sophisticated TGGE apparatus have been designed to help maintain a constant temperature gradient, DGGE is more often chosen as the preferred technique.

Terminal restriction fragment length polymorphism

Terminal restriction fragment length polymorphism (T-RFLP) has also been used to analyze microbial communities. This fingerprinting technique permits an automated quantification of the fluorescence signal intensities of the individual terminal restriction fragments in a given community fingerprint pattern (Lukow et al., 2000; Kent and Triplett, 2002). It is based on restriction endonuclease digestion of fluorescently end-labeled PCR products. The digested products are then separated by gel electrophoresis, which is detected on an automated sequence analyer. Therefore, depending on the species composition of the community, a varying fingerprint will be obtained (Lukow et al., 2000). This can offer phylogenetic information directly without further sequencing of the fragments, since the terminal restriction fragment lengths obtained can be compared to known 16S rRNA gene sequences (Kent and Triplett, 2002). However, when dealing with complex environmental samples, the T-RFLP profiles can be very complicated and hinder phylogenetic assignment of individual fragments as well as make comparisons and comprehension of the data quite difficult. This would then require the more tedious, and expensive, process of constructing a clone library from the amplified markers.

Clone libraries

Clone libraries can offer extremely good coverage of total bacterial identification, and an immense amount of information with full-length sequences. This pioneering method of bacterial community identification is, however, faulted by its very high cost due to the randomness of the identification. By randomly cloning PCR amplicons at near-full-length sequences, one can eventually obtain a good representation of the bacterial population in the sample; however, this will require sequencing several hundreds of clones, many of which will end up being the same dominant bacteria time and again (Brooks et al., 2003; Gonzalez et al., 2003). This wastes both time and money when trying to characterize biofilms. By combining the power of a clone library with DGGE, one can cut down on repetitive sequencing, and therefore reduce the cost of this procedure and increase the identification to full-length sequencing

(Gonzalez et al., 2003; Burr et al., 2006). Coupling of clone libraries to automated ribosomal intergenic spacer analysis (ARISA) has also been used to show microbial diversity (Brown et al., 2005).

Gene chips

DNA chip or microarray technology is becoming more widely used as an emerging method to identify and enumerate bacterial pathogens. This technology has had a major impact on identification of specific bacterial species within a wide field of study (Kent and Triplett, 2003; Ashbolt, 2003). The major advantage of this technology is that once a gene chip is created that includes all the necessary target probes, running the procedure to identify bacterial species within a sample is theoretically very robust, reliable, and rapid. This procedure can also very quickly and accurately determine changes in the population when comparing different samples and, depending on the set up, could even provide some information on relative quantities. Unfortunately, this technique at the moment is still extremely expensive, very laborious, and currently not very sensitive, according to Ashbolt (2003). It is also very time-consuming and difficult to set up and create your own chips. The major disadvantage is that the target sequence must be known, and therefore cannot be used to identify unknowns in the sample. Thus, although there is a lot of promise for gene chips when searching for a suite of specific microorganisms, the information that can be obtained about biofilm composition is relatively limited.

Peptide nucleic acid probes and fluorescence in situ hybridization

Peptide nucleic acid (PNA) probes have also emerged as a rapid technique for bacterial identification. PNA molecules are DNA mimics in which the negatively charged sugar-phosphate backbone of DNA is replaced with a noncharged polyamide or peptide backbone. These are much more useful than DNA probes since they are much smaller, and the noncharged backbone does not encounter electrostatic repulsion, allowing them to hybridize rapidly and tightly to nucleic acid targets (Stender et al., 2002; Lehtola et al., 2006; Wilks and Keevil, 2006). When combined with identification assays, such as fluorescence *in situ* hybridization (FISH), confirmation of a specific target is rapid and accurate (Lehtola et al., 2006, Wilks and Keevil, 2006). However, these probes are specifically designed and can therefore only be used to identify targeted species. FISH with DNA or PNA probes is suitable as a quick technique for checking bacterial species of interest, but FISH with DNA probes often suffers from significant background fluorescence when performed directly on biofilms; such interference is much less of a problem when PNA probes are used (Lehtola et al., 2006; Wilks and Keevil, 2006).

Who does what?

The techniques described above were initially developed and used to answer the question: Who (which bacteria or other microorganisms) are there in natural assemblages such as biofilms? An equally or more important question from the point of view of functional communities is: Which bacteria or other microorganisms perform what role in the community? To some extent, this can be answered by using the same techniques before and after challenge of a community with an appropriate intervention and looking for the responders (or nonresponders). However, another technique worth mentioning is achieving a certain level of prominence in understanding microbial community function. This has come about because of the comparatively recent coupling of molecular biological methods, such as those described, with analysis of stable isotope abundance (Radajewski et al., 2000; Ginige et al., 2004; Tillman et al., 2005; Wagner, 2005; Staal et al., 2007) and the greater availability of stable isotope precursors. It provides a culture-independent method of linking the identity of bacteria with their environmental function and specific metabolic processes using close to *in situ* conditions and will continue to be used widely in microbial ecology, including investigations into biofilms. It may be used to go beyond the effects of disinfection on the types of microorganisms present following disinfection, to determining the effects that disinfectants have on microbial metabolism under defined conditions.

Disinfection influences

When changes in drinking water disinfection regimes are made, there has been little or no prior consideration of their downstream microbiological impacts. The only microbiological consideration is whether the delivered dose (concentration and contact time, or CT) is adequate to inactivate the regulated suite of pathogens at the treatment plant. Clearly, however, the type and concentration of delivered disinfectant have broader influences on drinking water microbiology than simply dealing with these pathogens. Although all types of microorganisms can theoretically escape disinfection by one means or another, broadly speaking, the nature and concentration of disinfectant encountered selects the range of microorganisms available for downstream biofilm formation. Inevitably, it will always be the least susceptible microorganisms that survive disinfection, and among those, it will be the fittest (ecologically speaking) that survive to dominate the biofilm communities based on the resources available in time and space.

Examination of disinfection influences on biofilm constituents and composition has been relatively limited, and the discussion of this topic is inevitably restricted to certain aspects of practical significance to drinking water.

A recent trend in disinfection of drinking water throughout North America is a switch from chlorine as a secondary disinfectant to chloramine. For this there is usually a restricted contact with chlorine before the addition of ammonia to form monochloramine. The driver for this switch is regulation of the levels of certain disinfection by-products created during treatment, and a demonstrated reduction of these levels when the disinfectant residual is monochloramine rather than free chlorine. There have been two primary observations that have triggered some of the drinking water biofilm studies that have been undertaken. The first of these was an apparent reduction in the occurrence of *Legionella* species isolated from water treated with mono-chloramine, and the second was the seasonal occurrence of nitrification epi-sodes in drinking water that had been treated with monochloramine—the first apparently beneficial and the second apparently detrimental.

Eichler et al. (2006) studied the bacterial community dynamics of a drinking water from source to tap, but using only water and not biofilm samples. They found, unsurprisingly, marked effects of disinfection on the microbial profiles obtained that were more pronounced for the RNA-based fingerprints. It appeared as though chlorination may promote nitrifying bacteria, but the study was not definitive on this count. The source water clearly seemed to influence the water delivered to the consumer tap.

More frequently, studies have used model systems to examine biofilm structure. Martiny and colleagues (2005) conducted a longitudinal study in a nonchlorinated model system looking at composition of biofilms and water for periods from 2 weeks to 3 years. Nitrite oxidizers were found in all samples, and the authors suggested that nitrate may be an impor-tant substrate in view of the low assimilable organic carbon. They further suggest that while microbial communities in drinking water may control against nitrite, they also contain many indigenous heterotrophs capable of depleting disinfectants such as chloramine. Hoefel et al. (2005) used culture-independent techniques to detect bacteria associated with loss of chloramine and found that ammonia-oxidizing bacteria were present along with others that may be active in nitrogen cycling; however, these investigators sampled only bulk water and not biofilm. Ammonia- and nitrite-oxidizing bacteria have also been found in pilot-scale chloram-inated distribution systems (Regan et al., 2002) and tend to occur season-ally (Wolfe et al., 1990).

Studies of an unfiltered municipal drinking water specifically tar-geted at the period during switching from chlorine to monochloramine (Pryor et al., 2004) showed that prior to the switch, some of the drink-ing water distribution system biofilms had been dominated by legionel-lae based on PCR-DGGE analyses. However, after the switch, the DGGE analysis showed an absence of the previous *Legionella* bands and an increase in those associated with mycobacteria, which became among the dominant organisms detected by PCR-DGGE. These observations were supported by isolations from consumer outlets in the distribution system (Pryor et al., 2004; Moore et al., 2006). The reduction in legionellae associated with use of monochloramine has now been confirmed in a separate study (Flannery et al., 2006). Although supporting study for the increase in mycobacteria has not been done, mycobacteria are ubiquitous, important human pathogens and highly resistant to disinfection; it is extremely likely that they will benefit from the reduction of competition brought about by disinfection. It is worthy of note that they were isolated in a high proportion of disinfected samples but not in those experiencing biological treatment (Norton and LeChevallier, 2000).

Although the overall reduction in legionellae may have some human health benefits, the switch to chloramine is clearly not a panacea for all utilities. At the utility in Florida, where the study showed an increase in mycobacteria, other problems with the switch to chloramine have included taste and odor complaints, episodes of nitrification in storage tanks, rapid clogging of consumer point-of-use filters with cyanobacteria and sulfur bacteria, and fecal coliform occurrences in the distribution system that were not seen with chlorine as the disinfectant. Thus, it is important to examine the system as a whole and not to assume that a decrease in one opportunistic pathogen is necessarily the answer.

Only two studies have characterized microbial populations in chlorinated and chloraminated water specifically to examine general differences in microbial communities (Williams et al., 2004; Gagnon et al., 2007). Williams et al. (2004) used a single pipe loop and direct sequence of a 16S rDNA clone library from water only isolated on specific days. Gagnon et al. (2007) used a range of conditions in annular reactors as well as PCR-DGGE of the extracted DNA from biofilm, and water samples as well as DNA extracted from R2A-cultures. For each disinfectant (Gagnon et al., 2007), analyses were performed in the absence of disinfection as well as at two relevant disinfectant concentrations. The approach of the two studies was somewhat different in that Williams et al. (2004) focused on grouping the sequences into different classes, whereas Gagnon et al. (2007) were deliberately trying to target similarities and differences between the treatments. Each concluded that the disinfectant treatment could have affected the microbial community composition, but neither used a real full-scale treatment facility, and both used the treatments in sequence rather than in parallel. Thus, the results, while strongly suggestive of differences, are not definitive. It has also been shown that certain bacteria apparently dominant in biofilms under chlorine were totally absent when chloramine was the disinfectant, which may suggest some different chemistry at the biofilm surface (Gagnon et al., 2007; Matias et al., 2006). Another observation worthy of mention is that as species disappeared at higher concentrations of disinfectant, other previously undetected sequences were seen (Gagnon et al., 2007). This suggests that competition plays a distinct role in biofilm composition, with bacteria readily available to fill any voids created by the loss of less resistant bacteria. The disinfectant concentrations used were only those relevant to drinking water. Although studies of this type are rare, another study compared the natural biofilms formed with chlorine dioxide and UV disinfection (Schwartz et al., 2003).

When studying biofilms and biofilm composition, it is important to be cognizant of the heterogeneity intrinsic to biofilm structure (Wimpenny et al., 2000), and that organisms found in one spot may not necessarily be replicated in another. Therefore, the question arises as to how many samples, or rather, how big a surface area is necessary to get a representative sample. Our studies suggest (Gagnon et al., 2007) that for the more dominant organisms, even small surfaces such as those in an annular reactor may be adequate because repeat coupons or water samples would frequently yield the same or similar species. What is uncertain is whether a larger surface area would necessarily yield the rarer species since the dominance of certain species may be paramount and always make it difficult to detect the full range of species present. However, as we become more familiar with the types of microorganisms that could be present, and perhaps target genes that are more specific to certain groups, potentially with a whole genome preamplification as suggested by Gonzalez et al. (2005), then we may be able to tease out more of the minor community components that could still play a significant role under specified circumstances.

It is important also to note that although dominant presence is likely to have equated with some past or current activity, it does not necessarily imply that these organisms are all currently active (Keinanen-Toivola et al., 2006). On the other hand, the rapidity with which a biofilm population can change in response to different substrates suggests high malleability and responsiveness among biofilm communities, and this is supported in that many of the clones identified as active corresponded with those characterized in a DNA-based clone library (Keinanen-Toivola et al., 2006).

Opportunities for the future

Obviously, the needs of the drinking water industry are to produce safe water for consumers to drink (Szewzyk et al., 2000). There is considerable evidence that bacteria of public health significance that can gain access into the distribution system are able to persist there in biofilms with or without the assistance of protozoa (Brown and Barker, 1999). The opportunities provided by culture-independent techniques, sometimes in conjunction with culture, to explore drinking water, drinking water biofilms, and the capabilities of various treatments, including biological treatment, will help guide developments in the drinking water industry in future years. Currently the resources and skills needed to perform much of this work make such analyses prohibitive for most drinking water utilities and tend to restrict them to research laboratories, but that may change in future years as more technologies become available and more processes become automated.

References

Albert-Weissenberger, C., Cazalet, C., and Buchreiser, C. 2007. *Legionella pneumophila*—A human pathogen that co-evolved with freshwater protozoa. *Cell. Mol. Life Sci.* 64:432–48.

Allen, M.J., Edberg, S.C., and Reasoner, D.J. 2004. Heterotrophic plate count bacteria—What is their significance in drinking water? *Int. J. Food Microbiol.* 92:265–74.

Amann, R.I., Ludwig, W., and Schleifer, K.-H. 1995. Phylogenetic identification and in situ detection of individual microbial cells without cultivation. *Microbiol. Rev.* 59:142–69.

Ashbolt, N.J. 2003. Methods to identify and enumerate frank and opportunistic bacterial pathogens in water and biofilms. In *WHO: Heterotrophic plate counts and drinking-water safety*, ed. J. Bartram, J. Cotruvo, M. Exner, C. Fricker, and A. Glasmacher, chap. 9, 146–76. London: IWA Publishing.

Azevedo, N.F., Pachecho, A.P., Keevil, C.W., and Vieira, M.J. 2006. Adhesion of water stressed *Helicobacter pylori* to abiotic surfaces. *J. Appl. Microbiol.* 101:718–24.

Azevedo, N.F., Vieira, M.J., and Keevil, C.W. 2003. Establishment of a continuous model system to study *Helicobacter pylori* survival in potable water biofilms. *Water Sci. Technol.* 47:155–60.

Baker, G.C., Smith, J.J., and Cowan, D.A. 2003. Review and re-analysis of domain-specific 16S primers. *J. Microbiol. Methods* 55:541–55.

Bartram, J., Cotruvo, J., Dufour, A., Hazan, S., and Tanner, B. 2004. Heterotrophic plate count IJFM introduction. *Int. J. Food Microbiol.* 92:239–40.

Batté, M., Appenzeller, M.R., Grandjean, D., Fass, S., Gauthier, V., Jorand, F., Mathieu, L., Boualam, M., Saby, S., and Block, J.C. 2003. Biofilms in drinking water distribution systems. *Rev. Environ. Sci. Biol. Technol.* 2:147–68.

Böckelmann, U., Janke, A., Kuhn, R., Neu, T.R., Wecke, J., Lawrence, J.R., and Szewzyk, U. 2006. Bacterial extracellular DNA forming a defined network-like structure. *FEMS Microbiol. Lett.* 262:31–38.

Boe-Hansen, R., Martiny, A.C., Arvin, E., and Albrechtsen, H.-J. 2003. Monitoring biofilm formation and activity in drinking water distribution networks under oligotrophic conditions. *Water Sci. Technol.* 47:91–97.

Braid, M.D., Daniels, L.M., and Kitts, C.L. 2003. Removal of PCR inhibitors from soil DNA by chemical flocculation. *J. Microbiol. Methods* 52:389–93.

Brooks, S.P.J., McAllister, M., Sandoz, M., and Kalmokoff, M.L. 2003. Culture-independent phylogenetic analysis of the faecal flora of the rat. *Can. J. Microbiol.* 49:589–601.

Brown, M.R.W., and Barker, J. 1999. Unexplored reservoirs of pathogenic bacteria: Protozoa and biofilms. *Trends Microbiol.* 7:46–50.

Brown, M.V., Schwalbach, M.S., Hewson, I., and Fuhrman, J.E. 2005. Coupling 16S-ITS rDNA clone libraries and automated ribosomal intergenic spacer analysis to show marine microbial diversity: Development and application to a time series. *Environ. Microbiol.* 7:1466–479.

Burr, M.D., Clark, S.J., Spear, C.R., and Camper, A.K. 2006. Denaturing gradient gel electrophoresis can rapidly display the bacterial diversity contained in 16S rDNA clone libraries. *Microbial Ecol.* 51:479–86.

Davey, M.E., and O'Toole, G.A. 2000. Microbial biofilms: From ecology to molecular genetics. *Microbiol. Mol. Biol. Rev.* 64:847–67.

Eichler, S., Christen, R., Höltje, C., Westphal, P., Bötel, J., Brettar, I., Mehling, A., and Hofle, M.G. 2006. Composition and dynamics of bacterial communities of a drinking water supply system as assessed by RNA- and DNA-based 16SrRNA gene fingerprinting. *Appl. Environ. Microbiol.* 72:1858–72.

Emtiazi, F., Schwartz, T., Marten, S.M., Krolla-Sidenstein, P., and Obst, U. 2004. Investigation of natural biofilms formed during the production of drinking water from surface water embankment filtration. *Water Res.* 38:1197–206.

Farleitner, A.H., Zibuschka, F., Burtscher, M.M., Lindner, G., Reischer, G., and Mach, R.L. 2004. Eubacterial 16S-rDNA amplicon profiling: A rapid technique for comparison and differentiation of heterotrophic plate count communities from drinking water. *Int. J. Food Micribiol.* 92:333–45.

Flannery, B., Gelling, L.B., Vugia, D.J., Weintraub, J.M., Salerno, J.J., Conroy, M.J., Stevens, V.A., Rose, C.E., Moore, M.R., Fields, B.S., and Besser, R.E. 2006. Reducing *Legionella* colonization of water systems with monochloramine. *Emerg. Infect. Dis.* 12:588–95.

Gagnon, G.A., Murphy, H.M., Rand, J.L., Payne, S.J., Springthorpe, S., Matias, F., Nokhbeh, R., and Sattar, S.A. 2007. *Coliforms in distribution systems: Integrated disinfection and antimicrobial resistance.* AwwaRF Report for Project 3087, AWWA Research Foundation, Denver, CO.

Garcia, T.M., Jones, S., Pelaz, C., Millar, R.D., and Abu Kwaik, Y. 2007. Acanthamoeba polyphaga resuscitates viable but non-culturable *Legionella pneumophila* after disinfection. *Environ. Microbiol.* 9:1267–77.

Gelsomino, A., Keijzer-Wolters, A.C., Cacco, G., and Dirk van Elsas, J. 1999. Assessment of bacterial community structure in soil by polymerase chain reaction and denaturing gradient gel electrophoresis. *J. Microbiol. Methods* 38:1–15.

Ginige, M.P., Hugenholtz, P., Daims, H., Wagner, M., Keller, J., and Blackall, L.L. 2004. Use of stable-isotope probing, in situ hybridization-microautoradiography to study a methanol-fed denitrifying microbial community. *Appl. Environ. Microbiol.* 70:588–96.

Gonzalez, J.M., Ortiz-Martinez, A., Gonzalez-delValle, M.A., Liaz, L., and Saiz-Jimenez, C. 2003. An efficient strategy for screening large cloned libraries of amplified 16S rDNA sequences from complex environmental communities. *J. Microbiol. Methods* 55:459–63.

Gonzalez, J.M., Portillo, M.C., and Saiz-Jimenez, C. 2005. Multiple displacement amplification as a pre-polymerase chain reaction (pre-PCR) to process difficult to amplify samples and low copy number sequences from natural environments. *Environ. Microbiol.* 7:1024–28.

Hilbi, H., Weber, S.S., Ragaz, C., Nyfeler, Y., and Urwyler S. 2007. Environmental predators as models for bacterial pathogenesis. *Environ. Microbiol.* 9:563–75.

Hoefel, D., Monis, P.T., Grooby, W.L., Andrews, S., Saint, C.P. 2005. Culture-independent techniques for rapid detection of bacteria associated with loss of chloramine residual in a drinking water system. *Appl. Environ. Microbiol.* 71:6479–88.

Holben, W.E., and Harris, D. 1995. DNA-based monitoring of total bacterial community structure in environmental samples. *Mol. Ecol.* 4:627–31.

Huws, S.A., McBain, A.J., and Gilbert, P. 2005. Protozoan grazing and its impact upon population dynamics in biofilm communities. *J. Appl. Microbiol.* 98:238–44.

Joubert, L.-M., Woolfaardt, G.M., and Botha, A. 2006. Microbial exopolymers link predator and prey in a model yeast biofilm system. *Microbial Ecol.* 52:187–97.

Juhas, M., Ebert, L., and Tümmler, B. 2005. Quorum sensing: The power of cooperation in the world of *Pseudomonas*. *Environ. Microbiol.* 7: 459–71.

Kadouri, D., and O'Toole, G.A. 2005. Susceptibility of biofilms to *Bdellovibrio bacteriovorus* attack. *Appl. Environ. Microbiol.* 71:4044–51.

Kalmbach, S., Manz, W., and Szewzyk, U. 1997. Dynamics of biofilm formation in drinking water: Phylogenetic affiliation and metabolic potential of single cells assessed by formazan reduction and in situ hybridization. *FEMS Microbiol. Ecol.* 22:265–79.

Keinanen-Toivola, M.M., Revetta, R.P., and Santo Domingo, J.W. 2006. Identification of active bacterial communities in a model drinking water biofilm system using 16SrRNA-based clone libraries. *FEMS Microbiol. Lett.* 257:182–88.

Kent, A.D., and Triplett, E.W. 2002. Microbial communities and their interactions in soil and rhizosphere ecosystems. *Annu. Rev. Microbiol.* 56:211–36.

King, C.H., Shotts, E.B., Wooley, R.E., and Porter, K.G. 1988. Survival of coliforms and bacterial pathogens within protozoa during chlorination. *Appl. Environ. Microbiol.* 54:3023–33.

Lawrence, J.R., Swerhone, G.D.W., Leppard, G.G., Araki, T., Zhang, X., West, M.M., and Hitchcock, A.P. 2003. Scanning transmission x-ray, laser scanning and transmission electron microscopy mapping of the exopolymeric matrix of microbial biofilms. *Appl. Environ. Microbiol.* 69:5543–54.

LeChevallier, M.W., Gullick, R.W., Karim, M., Friedman, M., and Funk, J.E. 2003. The potential for health risks from intrusion of contaminants into the distribution system from pressure transients. *J. Water Health* 1:3–14.

Lehtola, M.J., Torvinen, E., Miettinen, I.T., and Keevil, C.W. 2006. Fluorescence in situ hybridization using peptide nucleic acid probes for rapid detection of *Mycobacterium avium* subsp. *avium* and *Mycobacterium avium* subsp. *paratuberculosis* in potable-water biofilms. *Appl. Environ. Microbiol.* 72:848–53.

Lukow, T., Dunfield, P.F., and Liesack, W. 2000. Use of the T-RFLP technique to assess spatial and temporal changes in the bacterial community structure within an agricultural soil planted with transgenic and non-transgenic potato plants. *Microbiol Ecol.* 32:241–47.

Martiny, A.C., Albrechtson, H.-J., Arvin, E., and Molin, S. 2005. Identification of bacteria in biofilm and bulk water samples from a nonchlorinated model drinking water system: Detection of a large nitrite-oxidizing population associated with *Nitrospira* spp. *Appl. Environ. Microbiol.* 71:8611–17.

Matias, F.M.G., Springthorpe, S., Gagnon, G., and Sattar, S.A. 2006. How does the disinfection regime affect the microbial profile in downstream distribution biofilms? Paper presented at the Annual Conference of the Canadian Water Works Association, Saint John, New Brunswick.

Matz, C., and Kjelleberg, S. 2005. Off the hook—How bacteria survive protozoan grazing. *Trends Microbiol.* 13:302–7.

McBain, A.J., Bartolo, R.G., Catrenich, C.E., Charbonneau, D., Ledder, R.G., Rickard, A.H., Symmons, S.A., and Gilbert, P. 2003. Microbial characterization of biofilms in domestic drains and the establishment of stable biofilm microcosms. *Appl. Environ. Microbiol.* 69:177–85.

Moore, M.R., Pryor, M., Fields, B., Lucas, C., Phelan, M., and Besser, R.E. 2006. Introduction of monochloramine into a municipal water system: Impact on colonization of buildings by *Legionella* spp. *Appl. Environ. Microbiol.* 72:378–83.

Muyzer, G. 1999. DGGE/TGGE: A method for identifying genes from natural ecosystems. *Curr. Opin. Microbiol.* 2:317–22.

Muyzer, G., Brinkhoff, T., Nübel, U., Santegoeds, C., Schäfer, H., and Wawer, C. 2004. Denaturing gradient gel electrophoresis (DGGE) in microbial ecology. In *Molecular microbial ecology manual*, 2nd Edition, Kowalchuk, G.A., de Bruijn, F.J., Head, I.M., Akkermans, A.D., and van Elsas, J.D. eds. 3.13 New York: 743–770, Springer.

Muyzer, G., De Waal, E., and Uitterlinden, A.G. 1993. Profiling of complex microbial populations by denaturing gradient gel electrophoesis analysis of polymerase chain reaction-amplified genes coding for 16S rRNA. *Appl. Environ. Microbiol.* 59:695–700.

Muyzer, G., and Smalla, K. 1998. Application of denaturing gradient gel electrophoresis (DGGE) and temperature gradient gel electrophoresis (TGGE) in microbial ecology. *Antonie van Leeuwenhoek.* 73:127–41.

Norton, C.D., and LeChevallier, M.W. 2000. A pilot study of bacteriological population changes through potable water treatment and distribution. *Appl. Environ. Microbiol.* 66:2686–276.

O'Toole, G., Kaplan, H.B., and Kolter, R. 2000. Biofilm formation as microbial development. *Annu. Rev. Microbiol.* 54:49–79.

Parsek, M.R., and Greenberg, E.P. 2005. Sociomicrobiology: The connections between quorum sensing and biofilms. *Trends Microbiol.* 13:27–33.

Penna, V.T.C., Martins, S.A.M., and Mazzola, P.G. 2002. Identification of bacteria in drinking and purified water during the monitoring of a typical water purification system. *BMC Public Health* 2:13–24.

Pryor, M., Springthorpe, S., Riffard, S., Brooks, T., Huo, Y., Davis, G., and Sattar, S.A. 2004. Investigation of opportunistic pathogens in municipal drinking water under different supply and treatment regimes. *Water Sci. Technol.* 50:83–90.

Radajewski, S., Ineson, P., Parekh, N.R., and Murrell, J.C. 2000. Stable isotope probing as a tool in microbial ecology. *Nature* 403:646–49.

Reading, N.C., and Sperandio, V. 2006. Quorum sensing: The many languages of bacteria. *FEMS Microbiol. Lett.* 254:1–11.

Regan, J.M., Harrington, G.W., and Noguera, D.R. 2002. Ammonia- and nitrite-oxidizing bacterial communities in a pilot-scale chloraminated drinking water distribution system. *Appl. Environ. Microbiol.* 68:73–81.

Salcher, M.M., Hofer, J., Horňák, K., Jezbera, J., Sonntag, B., Vrba, J., Šimek, K., and Posch, T. 2006. Modulation of microbial predator-prey dynamics by phosphorus availability: Growth patterns and survival strategies of bacterial phylogenetic clades. *FEMS Microbial Ecol.* 60:40–50.

Schwartz, T., Hoffman, S., and Obst, U. 2003. Formation of natural biofilms during chlorine dioxide and UV disinfection in a public drinking water distribution system. *J. Appl. Microbiol.* 95:591–601.

Shapiro, J.A. 1998. Thinking about bacterial populations as multicellular organisms. *Annu. Rev. Microbiol.* 52:81–104.

Shiner, E.K., Rumbaugh, K.P., and Williams, S.C. 2005. Interkingdom signaling: Deciphering the language of acyl homoserine lactones. *FEMS Microbiol. Rev.* 29:935–47.

Skraber, S., Schijven, J., Gantzer, C., and deRoda Husman, A.M. 2005. Pathogenic viruses in drinking water biofilms: A public health risk? *Biofilms* 2:105–17.

Snelling, W.J., McKenna, J.P., Lecky, D.M., and Dooley, J.S.G. 2005a. Survival of *Campylobacter jejuni* in waterborne protozoa. *Appl. Environ. Microbiol.* 71:5560–71.

Snelling, W.J., Moore, J.E., McKenna, J.P., Lecky, D.M., and Dooley, J.S.G. 2005b. Bacterial-protozoa interactions; an update on the role these phenomena play towards human illness. *Microbes Infect.* 8:578–87.

Staal, M., Thar, R., Kuhl, M., van Loosdrecht, M.C.M., Wolf, G., de Brouwer, J.F.C., and Rjlstenbll, J.W. 2007. Carbon isotope fractionation in developing natural phototrophic biofilms. *Biogeosci. Discuss.* 4:69–98.

Stender, H., Fiandaca, M., Hyldig-Nielsen, J.J., and Coull, J. 2002. PNA for rapid microbiology. *J. Microbiol. Methods* 48:1–17.

Stoodley, P., Sauer, K., Davies, D.G., and Costerton, J.W. 2002. Biofilms as complex differentiated communities. *Annu. Rev. Microbiol.* 56:187–209.

Szewzyk, U., Szewzyk, R., Manz, W., and Schleifer, K.-H. 2000. Microbiological safety of drinking water. *Annu. Rev. Microbiol.* 54:81–127.

Thomas, V., Bouchez, T., Nicolas, V., Robert, S., Loret, J.F., and Lévi, Y. 2004. Amoebae in domestic water systems: Resistance to disinfection treatments and implication in *Legionella* persistence. *J. Appl. Microbiol.* 97:950–63.

Tillman, S., Strompl, C., Timmis, K.N., and Abraham, W.R. 2005. Stable isotope probing reveals the dominant role of *Burkholderia* species in aerobic degradation of PCBs. *FEMS Microbiol. Ecol.* 52:207–17.

Vincke, E., Boon, N., and Verstraete, W. 2001. Analysis of the microbial communities on corroded concrete sewer pipes—A case study. *Appl. Microbiol. Biotechnol.* 57:776–85.

Wagner, M. 2005. The community level: Physiology and interactions of prokaryotes in the wilderness. *Environ. Microbiol.* 7:483–85.

Wang, G.C.-Y., and Wang, Y. 1997. Frequency of formation of chimeric molecules as a consequence of PCR coamplification of 16S rRNA genes from mixed bacterial communities. *Appl. Environ. Microbiol.* 63:4645–50.

Watanabe, K., Kodama, Y., and Harayama, S. 2001. Design and evaluation of PCR primers to amplify bacterial 16S ribosomal DNA fragments used for community fingerprinting. *J. Microbiol. Methods* 44:253–62.

Werker, A.G., and Hall, E.R. 2001. Quantifying population dynamics based on community structure fingerprints extracted from biosolids samples. *Microb. Ecol.* 41:195–209.

Wilks, S.A., and Keevil, C.W. 2006. Targeting species-specific low-affinity 16S rRNA binding sites by using peptide nucleic acids for the detection of legionellae in biofilms. *Appl. Environ. Microbiol.* 72:5453–62.

Williams, M.M., Domingo, J.W.S., Meckes, M.C., Kelty, C.A., and Rochon, H.S. 2004. Phylogenetic diversity of drinking water bacteria in a distribution system simulator. *J. Appl. Microbiol.* 96:954–64.

Wimpenny, J., Manz, W., and Szewzyk, U. 2000. Heterogeneity in biofilms. *FEMS Microbiol. Rev.* 24:661–71.

Wolfe, R.L., Lieu, N.I., Izaguirre, G., and Means, E.G. 1990. Ammonia oxidizing bacteria in a chloraminated distribution system: Seasonal occurrence, distribution, and disinfection resistance. *Appl. Environ. Microbiol.* 56:451–62.

Wotton, R.S. 2004. The ubiquity and many roles of exopolymers (EPS) in aquatic systems. *Scientia Marina* 68(Suppl. 1):13–21.

Wotton, R.S., and Preston, T.M. 2005. Surface films: Areas of water bodies that are often overlooked. *Bioscience* 55:137–45.

Zhang, L.-H., and Dong, Y.-H. 2004. Quorum sensing and signal interference: Diverse implications. *Mol. Microbiol.* 53:1563–71.

Zubkov, M.V., and Sleigh, M.A. 1999. Growth of amoebae and flagellates on bacteria deposited on filters. *Microb. Ecol.* 37:107–115.

chapter nine

Understanding the importance of biofilm growth in hot tubs

Darla M. Goeres

Contents

Communal bathing has existed since ancient times in many cultures. The Romans, in particular, were known for their opulent bath houses. Throughout time, the practice of public bathing has been associated with personal hygiene, medicinal treatments, creating a relaxing environment, and a social opportunity. Public bathing and bath houses were also associated with disease and promiscuity. The English word *stew*, meaning "brothel," originated from the French word for *bath house*.

Communal bathing was prominently practiced in the United States in the early twentieth century. Due to large social changes such as women's liberation and racial segregation, American society transitioned from largely public baths and swimming pools to more private pools. In the latter part of the twentieth century, the practice of communal bathing was reinvented in the United States with the creation of the portable hot tub. In 1956, the Jacuzzi brothers invented the first hydrotherapy pump that could be placed in a bath tub and was used to soothe arthritic muscles (www.jacuzzi.com/about/history.php). In the 1960s, unrelated to the Jacuzzi invention, people began constructing simple wooden hot tubs using wood from old redwood vats and discarded wooden wine barrels. For some of these wooden hot tubs, wood-fired heaters were used to warm the water that filled the tub.

The first modern portable hot tubs that incorporated jets and heating and filtration systems entered the market in the early 1970s. In 1975, McCausland and Cox (1975) published a case study titled *"Pseudomonas Infection Traced to a Motel Whirlpool,"* once again establishing the link between communal bathing and the potential for infection or disease. This chapter will focus on biofilms in residential portable hot tubs, also known as home spas or whirlpool spas.

Hot tub design, operation, use, and regulation

To understand the link between biofilm and hot tubs requires a brief digression into how hot tubs are built and operated. A successful hot tub incorporates engineering design with balanced water chemistry to handle the microbial and organic demand the bathers introduce while providing a pleasurable experience. In simplest terms, hot tubs are vessels that hold approximately 1,000 to 1,550 L of water that is heated, treated with disinfectants, filtered, and used for multiple bathing events. At first, portable hot tubs were built from scaled-down swimming pool components, until the industry recognized that hot tubs were a unique environment that required design decisions relevant to the operating conditions. Hot tub designs evolved to maximize relaxation for users, making them more efficient, ergonomic, and aesthetically pleasing.

Portable hot tubs are designed with several common features in the circulation system. A pump draws the water through a drain at the water's surface (a skimmer) to remove the large debris, through a filter; a heat exchanger warms the water to a specified temperature; and then the water returns to the hot tub. The liquid volume of the hot tub passes through this circulation process about every 30 min. The maximum recommended temperature for the water in a hot tub is 40°C. Hot tub cartridge filters are made from nonwoven polyester fabric pleated like an accordion around a piece of piping with holes drilled into it. The filter material is held in place with a top and bottom end cap. ANSI/NSF recommends that surface-type filters for use in residential hot tubs are sized with a maximum loading of 4.1 ml min^{-1} cm^{-2} (ANSI/NSF 50, 2001). Microscopic imaging has shown that bacteria are able to attach to the fibers of a hot tub filter (Goeres et al., 2007). Hot tub filters require regular cleaning because larger particles, such as hair, dead skin cells, or dirt, will also accumulate on the filter. Since pumps force water through the filter, filter pressure increases as the cartridge gets dirty. Manufacturers typically recommend that filters be cleaned when the pressure increases to a predetermined amount (typically 10 psi), and that cartridges be replaced annually.

Hot tubs contain a separate circulation system with air blowers and hydrotherapy jets, which help soothe sore muscles, increase blood flow to the central organs, provide respiratory exercise, and provide a relaxing effect (Becker, 1997). Besides their intended purpose, the air blowers and jets also aerate and mix the water. The number and configuration of the jets in a hot tub varies, depending upon the particular make and model of the hot tub. A typical residential hot tub is built to accommodate four to six bathers. The maximum number of bathers allowed per hot tub is calculated by dividing the surface area of the water in the tub by the 0.8361 m² allotted to each person (ANSI/APSP-2, 1999). It is not uncommon for hot tubs to operate at or even beyond design capacity, which negatively impacts the water quality because of the demand bathers exert on the disinfectant and the formation of disinfection by-products that results (Judd and Black, 2000; Kim et al., 2002).

Bathers introduce contaminants into the water in two ways. First, an initial large release of contaminants, including microorganisms, occurs when a person enters the water. These contaminants include skin, hair, body oils, lotions, minerals, and other organic substances on the body and swimsuit. The second type of contamination comes from the continuous release of sweat and dead skin cells into the water as the bather continues to soak. The contaminants then impact the water quality in two different ways. First, the microorganisms in the water can reproduce if no disinfectant residual is present. Second, if disinfectant is present, the contaminants may react with the disinfectant to lower the concentration until it is exhausted.

In practice, it is important to understand the different types of contamination when modeling the rate of disinfectant consumption in a hot tub. For example, consider the disappearance of chlorine during a bathing event. When bathers first enter the hot tub, the chlorine concentration will have an initial consumption rate associated with the large release of contaminants. A second, and most likely different, consumption rate will result as bathers continue to soak in the hot tub water. Of course, the water temperature, pH, whether the hydrotherapy jets are operating, and the impact of ultraviolet rays are also important to consider when modeling chlorine consumption.

Design and operational parameters that distinguish hot tubs from swimming pools, including high temperatures, heavy bather loads, aeration, large surface area-to-volume ratios, and different water turnover rates, elucidate the difficulty in maintaining balanced water quality in a hot tub. Hot tubs with poor water quality expose the user to a risk of infection, illness, or disease. As a result, draining the old water, cleaning the tub, and refilling it with fresh water is a good routine practice. Commonly, the recommended frequency, in days, for draining a hot tub is equal to 1/3(spa volume, U.S. gallons)/(maximum number of daily uses), or when the total dissolved solids (TDS) in the water exceed the source water TDS by 1,500 ppm (ANSI/APSP-2, 1999). In the interim, water quality is maintained through the addition of chemicals and filtration. The recreational water industry has established a set of guidelines for acceptable levels of oxidizing disinfectants, stabilizers, pH, alkalinity, calcium hardness, and TDS in hot tub water (ANSI/APSP-2, 1999; ANSI/APSP-3, 1999; ANSI/APSP-6, 1999).

Microorganisms enter a hot tub through various mechanisms. In addition to the microbial contamination that bathers introduce, microorganisms may enter the tub from the air, the supply water or garden hose used to fill the hot tub, dirt or other debris that may blow into an uncovered hot tub, contaminated toys, and contaminated hot tub covers. Similar to drinking water, hot tub water is not completely free of microorganisms. The goal is to keep the water sanitary to prevent the spread of disease.

In the United States the Environmental Protection Agency (EPA) is required by federal law (Federal Insecticide, Fungicide, and Rodenticide Act (FIFRA)) to register any product that makes pesticidal or antimicrobial claims; this includes hot tub disinfectants. Among several requirements, the agency requires that a disinfectant pass an efficacy test against both *Escherichia coli* and *Streptococcus faecalis* (AOAC, 1990) and a field test (EPA, 1979). The registration requirements outlined in the EPA document DIS/ TSS 12 were originally intended to verify the efficacy of swimming pool disinfectants. As disinfectants for hot tubs entered the market, the same registration requirements were applied to them, even though hot tubs operate at much different conditions than swimming pools. Other authors have acknowledged these differences between hot tubs and swimming pools and stated the need for new standards developed specifically for hot tubs (Crandall and MacKenzie, 1984; Favero, 1984; Kush and Hoadley, 1980). To date, nothing has changed. It is important to note that the current registration documents only include measuring the efficacy of disinfectants against suspended microorganisms. Biofilm is not discussed for either hot tubs or swimming pools, even though the importance of biofilm in recreational water venues has been known since the 1980s (Favero, 1984; Highsmith et al., 1985).

Although chemical hot tub disinfectants are regulated at the federal level, compliance rules for bacterial contamination in hot tubs and enforcement of those rules are the responsibility of the federal government and state and local health departments. Currently, states are not required to coordinate their activities in this area. This means that there are three different entities governing hot tub design, use, and water quality: self-regulation by the recreational water industry, a federal regulatory agency, and state and local health departments. In addition, a forth and separate entity, the Centers for Disease Control and Prevention (CDC), is responsible for tracking all waterborne disease outbreaks associated with hot tub use for the entire United States, an onerous task considering that most often infections that result from hot tub use go unreported. The most common hot tub infection, hot tub folliculitis, is self-limiting and does not require any medical attention. The people that use public hot tubs are often transient, as is the case at all hotel, motel, or cruise ship hot tubs. Doctors that treat the infection/illness may fail to either recognize a hot tub as the source of the infection or report the infection/illness to the health department. Normally, the infections that are reported are those that impact a number of people or have a serious illness or death associated with the outbreak. What this implies is that the real risk of infection or illness associated with hot tub use is not known.

Consider the challenge of maintaining a healthy hot tub due to the operating conditions alone. Balancing the water chemistry and maintaining a disinfectant residual in water with an elevated temperature, aeration, and heavy organic demand is possible, but requires a commitment from the hot tub owner or operator. In addition, hot tub disinfectants are regulated by the federal government using standards developed for a different system. These standards were put in place with no real data supporting the risk of infection associated with hot tub use. Compliance rules and enforcement are undertaken by state and local governments that do not necessarily communicate with each other or the federal agency regulating the disinfectants. Finally, many cases of hot-tub-acquired illness or infection go unreported. Considering all the above, it is not surprising that hot tubs fail to meet minimum water quality standards as often as they do. The next sections will discuss the importance of considering biofilm growth in hot tubs, adding an additional parameter to an already complex environment.

Hot tubs, biofilm, and waterborne disease

The literature contains many examples of illnesses associated with hot tub use, including a collection of "Surveillance for Waterborne Disease and Outbreaks—United States" reports issued by the U.S. Centers for Disease Control and Prevention (CDC) that date back to 1978 and report on recreation water disease outbreaks (http://www.cdc.gov/mmwr/). The published literature showed that the most common reported illnesses were those associated with ubiquitous opportunistic bacteria that can grow at high temperatures, such as *Pseudomonas*, *Legionella*, and *Mycobacteria* (Dziuban et al., 2006; Schulze-Röbbecke and Buchholtz, 1992). Other types of microorganisms have also been associated with hot tub disease (Cox et al., 1985; Holmes et al., 1989; Samples et al., 1984), but they account for fewer of the reported outbreaks.

Outbreaks due to *Pseudomonas aeruginosa* exposure causing folliculitis (skin rashes) are the most commonly reported (Chandrasekar et al.,

1984; Gustafson et al., 1983; McCausland and Cox, 1975; Molina et al., 1991; Ratnam et al., 1986; Spitalny et al., 1984b). The risk of developing hot tub folliculitis is dependent upon various factors, including the host's response to *P. aeruginosa* (Highsmith et al., 1985; Solomon, 1985) and the duration of time a bather soaked in the hot tub (Birkhead et al., 1987; Highsmith and Favero, 1985; Highsmith and McNamara, 1988; Hudson et al., 1985). Hot tub folliculitis is so common that many websites discuss this infection. Ear (Havelaar et al., 1983), eye (Insler and Gore, 1986), urinary (Salmen et al., 1985), and pneumonia infections (Crnich et al., 2003; Rose et al., 1983) are also associated with high levels of *P. aeruginosa* in hot tub water.

Respiratory illness is a potentially lethal threat to hot tub users or people who come in the vicinity of a contaminated hot tub. The literature contains examples of Legionnaires' disease (Jernigan et al., 1996; Vogt, 1987) or Pontiac fever (Fallon and Rowbotham, 1990; Fields et al., 2001; Götz et al., 2001; Mangione at al., 1985; Spitalny et al., 1984a) associated with using a hot tub contaminated with *Legionnella*. The bathers become infected when the *Legionnella* is aerosolized by the hot tub's jet and blower system. Whether bathers develop Legionnaires' disease, characterized as a severe pneumonia, or Pontiac fever, a self-limited influenza-like condition, depends upon their age and health (Thomas et al., 1993). In contrast to hot tub folliculitis, which is self-limiting, Legionnaires' disease is much more serious and may result in the bather's death.

Hot tub lung is another emerging disease reported in the literature that is associated with aerosolized nontuberculous mycobacteria (NTM), most often *Mycobacterium avium* complex (MAC) (Kahana et al., 1997; Lumb et al., 2004; Mangione et al., 2001). Although there is some debate whether to classify hot tub lung as an infection or hypersensitivity pneumonitis (Agarwal and Nath, 2006; Embil et al., 1997; Hanak, et al., 2005; Marchetti et al., 2004; Rickman et al., 2002), regardless, it is a serious illness that requires medical attention. For any respiratory illness associated with aerosolized bacteria, locating the hot tub indoors exasperates the problem (Schafer et al., 2003; Hanak et al., 2005). In addition to the respiratory disease outbreaks, a case of granulomatous dermatitis that resulted from *Mycobacterium intermedium* in a hot tub was reported in the literature (Edson et al., 2006).

Directly linking biofilm to infection, disease, or illness associated with hot tub use is extremely difficult. Biofilm is known to harbor pathogens, including the three species of bacteria discussed above (De Groote and Huitt, 2006; Hall-Stoodley and Lappin-Scott, 1998; Hall-Stoodley and Stoodley, 2005; Murga et al., 2001). Once a biofilm is established, it is well documented that biofilm bacteria are notoriously more tolerant to chemical disinfection than suspended microorganisms (Donlan and Costerton,

2002; Stewart et al., 2000). Biofilm can form in environmental systems even in the presence of low levels of disinfectants (Goeres et al., 2004, Pitts et al., 2001; Seyfried and Fraser, 1980), and bacteria begin expressing a biofilm phenotype within hours of becoming associated with a surface (Sauer and Camper, 2001). For hot tubs in particular, the difficulty of maintaining a constant disinfectant residual during periods of heavy use provides the bacteria the opportunity to form a biofilm on the hot tub surfaces (Dziuban et al., 2006).

The dramatic change in fluid shear that occurs when the jets and blowers are first turned on after a hot tub has not been in use for a period of time may encourage a biofilm sloughing event and the transmission of pathogens contained within the sloughed biofilm (Hall-Stoodley and Stoodley, 2005). Alternatively, the biofilm or individual bacteria can be transported in an aerosol and inhaled to reach the bather's lungs (Baron and Willeke, 1986). The detached biofilm retains a resistance to disinfection (Fux et al., 2004) and may contain hundreds of cells (Wilson et al., 2004), enough to be considered an infectious dose.

All of this is circumstantial evidence, though. Biofilm has not yet been directly linked to any hot tub disease outbreaks. Partly, this is because understanding the microbial ecology present during an outbreak is challenging. There is a lag time between a hot tub becoming contaminated, resulting in people becoming sick enough to visit a doctor, and the health department collecting samples to confirm the epidemiology of the outbreak. To complicate the situation further, the hot tub water is often shock treated with disinfectant during this lag time. Finally, when the testing does occur, biofilm samples are not always collected, especially from the piping, which is mostly inaccessible. Two examples of exceptions to the above statements include an outbreak in a swimming pool where biofilm samples (referred to as a "swab of the debris on the edge of the pool") were positive for *Pseudomonas aeruginosa*, even after a lag time and heavy chlorination had occurred (Gustafson et al., 1983), and in a hot tub where filter samples were positive for *Legionella micdadei* 25 days after the hot tub that was the source of the outbreak had been hyperchlorinated (Fallon and Rowbotham, 1990).

An initial link between hot tubs, disease, and biofilm was made in the CDC's "Surveillance for Waterborne Disease and Outbreaks Associated with Recreational Water—United States, 2003–2004" (Dziuban et al., 2006). Biofilm was mentioned in association with hot tubs, and a "lack of essential cleaning of spas to minimize biofilm build-up" was listed as a common theme derived from the outbreaks reported (Dziuban et al., 2006). What is needed are two studies designed to answer the following questions: What is the typical community profile and density of biofilm growing in maintained hot tubs? What is the health risk associated with bathing in a hot tub that contains this typical biofilm community for an immunocompetent adult?

Biofilm in the hot tub research literature

The majority of publications that include *hot tub* as a key word discuss illnesses associated with hot tub use. The other main topic in the hot tub literature is the publication of field studies. For these publications, investigators sampled hot tubs for the presence of various bacteria, including *Pseudomonas* (Hewitt, 1985; Jacobson et al., 1976; Kush and Hoadley, 1980; Price and Ahearn, 1988), *Pseudomonas* and *Staphylococcus* (Schiemann, 1985), *Mycobacteria* (Havelaar et al., 1985), and *Legionella* (Groothuis et al., 1985). Schiemann (1985) reported on the importance of sampling for bacteria in the surface film that floats on top of the water. Other papers discuss novel sampling procedures for detecting bacterial contamination (Johnson et al., 1994; Lee et al., 2001) or the potential for hot tub aerosols to transmit bacteria from the water to the air (Baron and Willeke, 1986; Blanchard and Syzdek, 1982; Mangione et al., 1985; Schafer et al., 2003). Field surveys confirmed that the majority of hot tubs sampled contained bacteria. Also, the hot tubs commonly had too little disinfectant and an improper pH level.

Few papers are published in the scientific literature that present basic or applied research results on hot tubs (Goeres et al., 2007; Highsmith et al., 1985; Nerurkar et al., 1983; Shaw, 1984; Watt et al., 2000). Fewer still focus on biofilms in hot tubs (Goeres et al., 2007), even though the importance of biofilm in hot tubs is not a novel idea (Dziuban et al., 2006; Chandrasekar et al., 1984; Favero, 1984; Highsmith et al., 1985; Price and Ahearn, 1988). One explanation for the low number of research publications is the limited funding available to explore the basic questions associated with biofilm in hot tubs. The outcome is that results published on swimming pool research are often extrapolated to hot tubs. Another likely explanation is that the recreational water industry is highly competitive. The research and development departments of the recreational water product manufacturing companies most likely have been studying biofilm in hot tubs for years, but often do not publish their findings. Consequently, a lot of effort is duplicated and money wasted researching questions that another company has already answered. The knowledge these companies possess is only apparent from analyzing the products and associated marketing strategy that the company uses to sell those products. Even though not much is published in the scientific literature, there are respected organizations, such as the CDC (www.cdc.gov/healthyswimming/index), National Swimming Pool Foundation (www.nspf.org), and World Health Organization (www.who.int/water_sanitation_health/bathing/en), that are dedicated to providing recreational water educational material.

Understanding the link between biofilm and suspended bacteria

Currently, hot tub water samples are collected to measure the sanitary condition. The water is neutralized, plated, and incubated before viable cells are counted (Eaton et al., 2005). If the bulk water is free of bacteria, then one concludes that the hot tub is acceptable for bathing. However, testing the water for bacteria was shown to be an inaccurate measure of the sanitary condition of a hot tub. Researchers at the Center for Biofilm Engineering correlated biofilm heterotrophic plate counts (HPCs) to bulk water HPCs collected from the same hot tub during the same sampling event. Various hot tubs in Bozeman, Montana, were sampled over a period of months. The bulk water samples were collected using the method described above. Biofilm samples were collected by laying a sterile strip of neoprene rubber that contained a 1.267 cm^2 hole as close to the skimmer inlet as possible. To collect the biofilm, the inside of the hole was scraped with a wooden applicator stick for approximately 15 s; then the stick was vigorously rinsed in 10 ml of sterile phosphate-buffered saline that contained an appropriate neutralizer. The scrape and rinse procedure was repeated three or four times at the same location for each biofilm sample. Both the bulk water and biofilm samples were brought back to the laboratory for further analysis. The biofilm sample was homogenized to disaggregate the clumps, diluted, spread plated on R2A agar, incubated at 35°C for 5 days, and the colonies enumerated. The bulk water sample was vigorously vortexed, diluted, spread plated, incubated at 35°C for 5 days, and the colonies enumerated.

Figure 9.1 shows a limited correlation between the bulk water and biofilm heterotrophic plate counts. The vertical line in Figure 9.1 represents a bulk water HPC of 200 cfu/ml (2.3 log_{10}cfu/ml), which is commonly set as the maximum level of bacteria acceptable in hot tub water. When the hot tub did not meet this standard, the biofilm counts were roughly correlated to the bulk water counts. But, when the bulk water counts were less than 200 cfu/ml, the biofilm counts ranged from less than the level of detection to slightly more than 4.5 log_{10}cfu/cm^2.

These data demonstrate that it is not possible to predict biofilm contamination based solely on bulk water counts. Therefore, standard microbiological tests are of little use when predicting if a biofilm is present within the tub and if that biofilm imparts an added risk to the bather. In 1984, Favero theorized the possibility of biofilm to serve as the source of contamination for hot tub water. In 1988, Price and Ahearn drew a similar conclusion when they attributed the recolonization of *P. aeruginosa* in hot tub water to the presence of biofilm. Thus, bacteria introduced during a bathing event may impart some bather risk; however, bacteria that survive in a biofilm also contribute to bather risk.

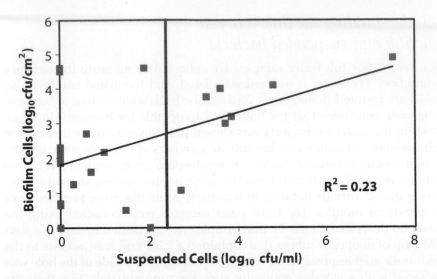

Figure 9.1 Minimal correlation was found between biofilm and planktonic heterotrophic bacteria recovered from field hot tub samples. Each point represents a paired biofilm-bulk water sample that was collected by one technician from a single hot tub. Samples were collected on the same day within 30 mins of each other. The vertical line represents 200 cfu/ml.

As a simple calculation shows, a 1,400 L hot tub with 200 cfu/ml has a total of 2.8×10^8 cfu present in the water. Now consider the bacteria associated with the surfaces in the hot tub, including the piping, filter, and hot tub shell. If this same hot tub has a surface area of 20 m^2 and an average biofilm accumulation of 1×10^3 cfu/cm^2, the total number of colonies associated with the surface is 2×10^8. The means that roughly 58% of the total bacteria in the hot tub are in the bulk water and 42% are associated with a surface. Said another way, the planktonic bacteria are roughly in balance with the biofilm bacteria. Of course, this hot tub would not be considered safe for bathing based on the 200 cfu/ml limit. If the same 1,400 L hot tub contained 10 cfu/ml, then the total colonies in the bulk water would equal 1.4×10^7. As Figure 9.1 demonstrates, the total number of biofilm bacteria in the hot tub could still equal 2×10^8, but now the planktonic population only contributes 7% of the total number of bacteria present in the hot tub and the biofilm is contributing 93% of the population. In this example, the bulk water sample is not an adequate representation of the bacterial risk to humans. Worse, one would conclude from the bulk water microbiological test that this hot tub is safe for bathing, while neglecting the risk associated with the biofilm bacteria. Understanding the importance of biofilm demonstrates the need for protocols that health officials can use to collect biofilm samples from hot tub surfaces, including the inaccessible piping, to determine the real health of the hot tub.

Biofilm control and removal

Biofilm may establish on every surface of the hot tub, including the hot tub shell, piping, and filters. Once a biofilm is established in a hot tub, it is extremely difficult to kill or remove (Storey, 1989). Draining the water and scrubbing will work for accessible surfaces. For inaccessible surfaces, biocides that are effective against biofilm are needed. Unfortunately, at this time, no standard methods, regulations, or even recommendations are in place to verify the efficacy of these treatments against biofilm.

Two cautionary notes: First, effectively removing biofilm from the surfaces of a contaminated hot tub may temporarily put bathers at an increased risk for infection, especially from bacteria that prefer to partisan as an aerosol (Angenent et al., 2005). Second, treatments such as UV, which do not leave a residual, are not effective against biofilm. Ozone may have limited effectiveness because a residual is not maintained in all locations that contain biofilm. That is why it is recommended that these types of treatments are used in combination with a chemical that maintains a residual, such as chlorine, bromine, or chlorine dioxide.

Conclusions and recommendations

Biofilm exists in hot tubs and may thrive even when the planktonic bacterial population is well below the acceptable level or not detected. Biofilm protects pathogenic bacteria associated with hot tub outbreaks. Based on the lack of data, evidence is circumstantial that biofilm contributes to outbreaks; therefore, protocols to assess the presence of biofilm in outbreak investigations are needed. Also needed is basic research to answer the following questions:

- What is the typical biofilm density on the surfaces of a hot tub that is considered well maintained?
- What species of microorganisms populate the typical hot tub biofilm?
- What is the risk of illness for healthy adults who use hot tubs populated with a typical biofilm when a disinfectant residual is maintained?

From the regulatory and enforcement perspective, what is needed are new registration guidelines developed specifically for hot tubs that include protocols to minimize the risk of exposure to biofilm bacteria.

Soaking in hot tubs is a health activity with cardiac, respiratory, and relaxation health benefits. These systems also pose a health risk if not properly maintained. Some ambiguity remains that must be resolved to ascertain the risk biofilm creates for users. Until the risk is thoroughly understood, diligence in maintaining a proper disinfectant level, maintaining balanced water chemistry, routinely cleaning or replacing the filter, draining and cleaning the hot tub, limiting the number of bathers to below the designed capacity, and requiring bathers to shower before entering will help reduce the risk biofilm can impart.

Acknowledgments

The author thanks Joanna Heersink and Linda Loetterle for their assistance in collecting and analyzing the hot tub field samples. The hot tub field research and literature review was supported by the U.S. Environmental Protection Agency (EPA) under contract 68-W-02-050 with Montana State University. Statements in this chapter do not necessarily reflect the views of the EPA.

References

Agarwal, R., and Nath, A., Hot-tub lung: Hypersensitivity to *Mycobacterium avium* but not hypersensitivity pneumonitis, *Respiratory Med.*, 100, 1478, 2006.

Angenent, L.T., Kelley, S.T., Amend, A.St., Pace, N.R., and Hernandez, M.T., Molecular identification of potential pathogens in water and air of a hospital therapy pool, *Proc. Natl. Acad. Sci. USA*, 102, 4860–65, 2005.

ANSI/APSP-2, *American national standard for public spas*, National Spa and Pool Institute, Alexandria, VA, 1999.

ANSI/APSP-3, *American national standard for permanently installed residential spas*, National Spa and Pool Institute, Alexandria, VA, 1999.

ANSI/APSP-6, *American national standard for portable spas*, National Spa and Pool Institute, Alexandria, VA, 1999.

ANSI/NSF 50, *Circulation system components and related materials for swimming pools, spas/hot tubs*, NSF International, Ann Arbor, MI, 2001.

AOAC, Disinfectants (water) for swimming pools, AOAC Official Method 985.13, in *AOAC official methods of analysis*, AOAC International, Gaithersburg, MD, 1990.

Baron, P.A., and Willeke, K., Respirable droplets from whirlpools: Measurements of size distribution and estimation of disease potential, *Environ. Res.*, 39, 8–18, 1986.

Becker, B.E., Biophysiologic aspects of hydrotherapy, in *Comprehensive aquatic therapy*, ed. B.E. Becker and A.J. Cole, Butterworth-Heinemann, Boston, 1997, 17–48.

Birkhead, G., Vogt, R.L., and Hudson, P.J., Whirlpool folliculitis, *Am. J. Public Health*, 77, 514, 1987.

Blanchard, D.C., and Syzdek, L.D., Water-to-air transfer and enrichment of bacteria in drops from bursting bubbles, *Appl. Environ. Microbiol.*, 43, 1001–5, 1982.

Chandrasekar, P.H., Rolston, K.V., Kannangara, D.W., LeFrock, J.L., and Binnick, S.A., Hot tub-associated dermatitis due to *Pseudomonas aeruginosa*: Case report and review of the literature, *Arch. Dermatol.*, 120, 1337–40, 1984.

Cox, G.F., Levy, M.L., and Wolf, J.E., Jr., Is eczema herpeticum associated with the use of hot tubs? *Pediatr. Dermatol.*, 2, 322–23, 1985.

Crandall, R.A., and MacKenzie, C.J., Pathogenic hazards and public spa *and* hot tub facilities, *Can. J. Public Health*, 75, 223–26, 1984.

Crnich, C.J., Gordon, B., and Andes, D., Hot tub–associated necrotizing pneumonia due to *Pseudomonas aeruginosa*, *Clin. Infect. Dis.*, 36, e55–57, 2003.

De Groote, M.A., and Huitt, G., Infections due to rapidly growing mycobacteria, *Emerg. Infect.*, 42, 1756–63, 2006.

Donlan, R.M., and Costerton, J.W., Biofilms: Survival mechanisms of clinically relevant microorganisms, *Clin. Microbiol. Rev.*, 15, 167–93, 2002.

Dziuban, E.J., Liang, J.L., Craun, G.F., Hill, V., Yu, P.A., Painter, J., Moore, M.R., Calderon, R.L., Roy, S.L., and Beach, M.J., Surveillance for waterborne disease and outbreaks associated with recreational water—United States, 2003–2004, *MMWR Surveillance Summaries*, 55(SS12), 1–24, 2006.

Eaton, A.D., Clesceri, L.S., Rice, E.W., and Greenberg, A.E., eds., *Standard methods for the examination of water and waste waters*, 21st ed., American Public Heath Association, American Water Works Association, Water Environment Federation, Washington, DC, 2005, 9213 C.

Edson, R.S., Terrell, C.L., Brutinel, W.M. and Wengenack, N.L., *Mycobacterium intermedium* granulomatous dermatitis from hot tub exposure, *Emerg. Infect. Dis.*, 12, 821–23, 2006.

Embil, J., Warren, P., Yakrus, M., Stark, R., Corne, S., Forrest, D., and Hershfield, E., Pulmonary illness associated with exposure to *Mycobacterium-avium* complex in hot tub water: Hypersensitivity pneumonitis or infection? *Chest*, 111, 813–17, 1997.

EPA, *Efficacy data requirements: Swimming pool water disinfectants*, DIS/TSS-12, www.epa.gov/oppad001/dis_tss_docs/dis-12.htm, April 23, 1979.

Fallon, R.J., and Rowbotham, T.J., Microbiological investigations into an outbreak of Pontiac fever due to *Legionella micdadei* associated with use of a whirlpool, *J. Clin. Pathol.*, 43, 479–483, 1990.

Favero, M.S., Whirlpool spa-associated infections: Are we really in hot water? *Am. J. Public Health* 74, 653–55, 1984.

Fields, B.S., Haupt, T., and Davis, J.P., Pontiac fever due to *Legionella micdadei* from a whirlpool spa: Possible role of bacterial endotoxin, *J. Infect. Dis.*, 184, 1289–92, 2001.

Fux, C.A., Wilson, S., and Stoodley, P., Detachment characteristics and oxacillin resistance of *Staphylococcus aureus* biofilm emboli in an in vitro catheter infection model, *J. Bacteriol.*, 186, 4486–91, 2004.

Goeres, D.M., Loetterle, L.R., and Hamilton, M.A., A laboratory hot tub model for disinfectant efficacy evaluation, *J. Microbiol. Methods*, 68, 184–92, 2007.

Goeres, D.M., Palys, T., Sandel, B.B., and Geiger, J., Evaluation of disinfectant efficacy against biofilm and suspended bacteria in a laboratory swimming pool model, *Water Res.*, 38, 3103–9, 2004.

Götz, H.M., Tegnell, A., De Jong, B., Broholm, K.A., Kuusi, M., Kallings, I., and Ekdahl, K., A whirlpool associated outbreak of Pontiac fever at a hotel in Northern Sweden, *Epidemiol. Infect.*, 126, 241–47, 2001.

Groothuis, D.G., Havelaar, A.H., and Veenendaal, H.R., A note on legionellas in whirlpools, *J. Appl. Bacteriol.*, 58, 479–81, 1985.

Gustafson, T.L., Band, J.D., Hutcheson Jr., R.H., and Schaffner, W., *Pseudomonas folliculitis*: An outbreak and review, *Rev. Infect. Dis.*, 5, 1–8, 1983.

Hall-Stoodley, L., and Lappin-Scott, H., Biofilm formation by the rapidly growing mycobacterial species *Mycobacterium fortuitum*, *FEMS Microbiol.*, 168, 77–84, 1998.

Hall-Stoodley, L., and Stoodley, P., Biofilm formation and dispersal and the transmission of human pathogens, *Trends Microbiol.*, 13, 7–10, 2005.

Hanak, V., Kalra, S., Aksamit, T.R., Hartman, T.E., Tazelaar, H.D., and Ryu, J.H., Hot tub lung: Presenting features and clinical course of 21 patients, *Respiratory Med.*, 100, 610–15, 2006.

Havelaar, A.H., Berwald, L.G., Groothuis, D.G., and Baas, J.G., Mycobacteria in semi-public swimming-pools and whirlpools, *Zentralbl. Bakteriol. Mikrobiol. Hyg. B*, 180, 505–14, 1985.

Havelaar, A.H., Bosman, M., and Borst, J., Otitis externa by *Pseudomonas aeruginosa* associated with whirlpools, *J. Hygiene Cambridge*, 90, 489–498, 1983.

Hewitt, J.H., *Pseudomonas aeruginosa* and whirlpools, *Br. Med. J. (Clin. Res. Ed.)*, 290, 1353–54, 1985.

Highsmith, A.K., and Favero, M.S., Microbiologic aspects of public whirlpools, *Clin. Microbiol. Newsl.*, 7, 9–11, 1985.

Highsmith, A.K., Le, P.N., Khabbaz, R.F., and Munn, V.P., Characteristics of *Pseudomonas aeruginosa* isolated from whirlpools and bathers, *Infect. Control*, 6, 407–12, 1985.

Highsmith, A.K., and McNamara, M., Microbiology of recreational and therapeutic whirlpools, *Toxicity Assessment*, 3, 599–611, 1988.

Holmes, S.E., Pearson, J.L., Kinde, M.R., and Hennes, R.F., Gastroenteritis outbreak: Disease linked to swimming pool and spa use, *J. Environ. Health*, 51, 286–88, 1989.

Hudson, P.J., Vogt, R.L., Jillson, D.A., Kappel, S.J., and Highsmith, A.K., Duration of whirlpool-spa use as a risk factor for *Pseudomonas dermatitis*, *Am. J. Epidemiol.*, 122, 915–17, 1985.

Insler, M.S., and Gore, H., *Pseudomonas keratitis* and *folliculitis* from whirlpool exposure, *Am. J. Ophthalmol.*, 101, 41–43, 1986.

Jacobson, J.A., Hoadley, A.W., and Farmer, J.J., *Pseudomonas aeruginosa* serogroup 11 and pool-associated skin rash, *Am. J. Public Health*, 66, 1092–93, 1976.

Jernigan, D.B., Hofmann, J., Cetron, M.S., Genese, C.A., Nuorti, J.P., Fields, B.S., Benson, R.F., Carter, R.J., Edelstein, P.H, Guerrero, I.C., Paul, S.M., Lipman, H.B., and Breiman, R., Outbreak of Legionnaires' disease among cruise ship passengers exposed to a contaminated whirlpool spa, *Lancet*, 347, 494–99, 1996.

Johnson, W.M., Bernard, K., Marrie, T.J., and Tyler, S.D., Discriminatory genomic fingerprinting of *Legionella pneumophila* by pulsed-field electrophoresis, *J. Clin. Microbiol.*, 32, 2620–21, 1994.

Judd, S.J., and Black, S.H., Disinfection by-product formation in swimming pool waters: A simple mass balance, *Water Res.*, 34, 1611–19, 2000.

Kahana, L.M., Kay, J.M., Yakrun, M.A., and Waserman, S., *Mycobacterium avium* complex infection in an immunocompetent young adult related to hot tub exposure, *Chest*, 111, 242–245, 1997.

Kim, H., Shim, J., and Lee, S., Formation of disinfection by-products in chlorinated swimming pool water, *Chemosphere* 46, 123–30, 2002.

Kush, B.J., and Hoadley, A.W., A preliminary survey of the association of *Pseudomonas aeruginosa* with commercial whirlpool bath waters, *Am. J. Public Health*, 70, 279–81, 1980.

Lee, J., Deininger, R.A., and Fleece, R.M., Rapid determination of bacteria in pools, *J. Environ. Health*, 64, 9–14, 2001.

Lumb, R., Stapledon, R., Scroop, A., Bond, P., Cunliffe, D., Goodwin, A., Doyle, R., and Bastian, I., Investigation of spa pools associated with lung disorders caused by *Mycobacterium avium* complex in immunocompetent adults, *Appl. Environ. Microbiol.*, 70, 4906–10, 2004.

Mangione, E.J., Huitt, G., Lenaway, D., Beebe, J., Bailey, A., Figoski, M., Rau, M.P., Albrecht, K.D., and Yakrus, M.A., Nontuberculous mycobacterial disease following hot tub exposure, *Emerg. Infect. Dis.*, 7, 1039–1042, 2001.

Mangione, E.J., Remis, R.S., Tait, K.A., McGee, H.B., Gorman, G.W., Wentworth, B.B., Baron, P.A., Hightower, A.W., Barbaree, J.M., and Broome, C.V., An outbreak of Pontiac fever related to whirlpool use, Michigan 1982, *JAMA*, 253, 535–39, 1985.

Marchetti, N., Criner, K., and Criner, G.J., Characterization of functional, radiologic and lung function recovery post-treatment of hot tub lung. A case report and review of the literature, *Lung*, 182, 271–77, 2004.

McCausland, W.J., and Cox, P.J., *Pseudomonas* infection traced to a motel whirlpool, *J. Environ. Health*, 37, 455–59, 1975.

Molina, D.N, Colon, M., Bermudez, R.H., and Ramirez-Ronda, C.H., Unusual presentation of *Pseudomonas aeruginosa* infections: A review, *Boletain de la Asociaciaon Muedica de Puerto Rico*, 83, 160–63, 1991.

Murga, R., Forster, T.S., Brown, E., Pruckler, J.M., Fields, B.S., and Donlan, R.M., Role of biofilms in the survival of *Legionella pneumophila* in a model potable-water system, *Microbiology*, 147, 3121–25, 2001.

Nerurkar, L.S., West, F., May, M., Madden, D.L., and Sever, J.L., Survival of herpes simplex virus in water specimens collected from hot tubs in spa facilities and on plastic surfaces, *JAMA*, 250, 3081–83, 1983.

Pitts, B., Willse, A., McFeters, G.A., Hamilton, M.A., Zelver, N., and Stewart, P.S., A repeatable laboratory method for testing the efficacy of biocides against toilet bowl biofilms, *J. Appl. Microbiol.*, 91, 110–17, 2001.

Price, D., and Ahearn, D. G., Incidence and persistence of *Pseudomonas aeruginosa* in whirlpools, *J. Clin. Microbiol.*, 26, 1650–54, 1988.

Ratnam, S., Hogan, K., March, S.B., and Butler, R.W., Whirlpool-associated folliculitis caused by *Pseudomonas aeruginosa*: Report of an outbreak and review, *J. Clin. Microbiol*, 23, 655–59, 1986.

Rickman, O.B., Ryu, J.H., Fidler, M.E., and Kalra, S., Hypersensitivity pneumonitis associated with *Mycobacterium avium* complex and hot tub use, *Mayo Clin. Proc.*, 77, 1233–37, 2002.

Rose, H.D., Franson, T.R., Sheth, N.K., Chusid, M.J., Macher, A.M., and Zeirdt, C.H., *Pseudomonas pneumonia* associated with use of a home whirlpool spa, *JAMA*, 250, 2027–29, 1983.

Salmen, P., Dwyer, D.M., Vorse, H., and Kruse, W., Whirlpool-associated *Pseudomonas aeruginosa* urinary tract infections, *JAMA*, 250, 2025–26, 1983.

Samples, J.R., Binder, P.S., Luibel, F.J., and Font, R.L., *Acanthamoeba keratitis* possibly acquired from a hot tub, *Arch. Ophthalmol.*, 102, 707–10, 1984.

Sauer, K., and Camper, A.K., Characterization of phenotypic changes in *Pseudomonas putida* in response to surface-associated growth, *J. Bacteriol.*, 183, 6579–89, 2001.

Schafter, M.P., Martinez, K.F., and Mathews, E.S., Rapid detection and determination of the aerodynamic size range of airborne mycobacteria associated with whirlpools, *Appl. Occupational Environ. Hygiene*, 18, 41–50, 2003.

Schiemann, D.A., Experiences with bacteriological monitoring of pool water, *Infection Control*, 6, 413–17, 1985.

Schulze-Röbbecke, R., and Buchholtz, K., Heat susceptibility of aquatic mycobacteria, *Appl. Environ. Microbiol.*, 58, 1869–73, 1992.

Seyfried, P.L., and Fraser, D.J., Persistence of *Pseudomonas aeruginosa* in chlorinated swimming pools, *Can. J. Microbiol.*, 26, 350–55, 1980.

Shaw, J.W., A retrospective comparison of the effectiveness of bromination and chlorination in controlling *Pseudomonas aeruginosa* in spas (whirlpools) in Alberta, *Can. J. Public Health*, 75, 61–68, 1984.

Solomon, S.L., Host factors in whirlpool-associated *Pseudomonas aeruginosa* skin disease, *Infect. Control*, 6, 402–6, 1985.

Spitalny, K.C., Vogt, R.L., Orciari, L.A., Witherell, L.E., Etkind, P., and Novick, L.F., Pontiac fever associated with a whirlpool spa, *Am. J. Epidemiol.*, 120, 809–17, 1984a.

Spitalny, K.C., Vogt, R.L., and Witherell, L.E., National survey on outbreaks associated with whirlpool spas, *Am. J. Public Health*, 74, 725–26, 1984b.

Stewart, P.S., McFeters, G.A., and Huang, C.-T., Biofilm control by antimicrobial agents, in *Biofilms II: Process analysis and applications*, ed. J.D. Bryers, Wiley-Liss, New York, 2000, 373–405.

Storey, A., Microbiological problems of swimming pools, *Environ. Health*, 97, 260–62, 1989.

Thomas, D.L., Mundy, L.M., and Tucker, P.C., Hot tub legionellosis: Legionnaires' disease and Pontiac fever after a point-source exposure to *Legionella pneumophila*, *Arch. Intern. Med.*, 153, 2597–99, 1993.

Vogt, R.L., Hudson, P.J., Orciari, L., Heun, E.M., and Woods, T.C., Legionnaire's disease and a whirlpool-spa, *Ann. Intern. Med.*, 107, 596, 1987.

Watt, P., Kennedy, D.I., Casanova, L.M., Sandoval, J., and Gerba, C.P., Healthy hot tubs and sickly spas, Paper Q-341 presented at the 100th General Meeting of the American Society for Microbiology, Washington DC, May 21–25, 2000.

Wilson, S., Hamilton, M.A., Hamilton, G.C., Schumann, M.R., and Stoodley, P., Statistical quantification of detachment rates and size distributions of cell clumps from wild-type (PAO1) and cell signaling mutant (JP1) *Pseudomonas aeruginosa* biofilms, *Appl. Environ. Microbiol.*, 70, 5847–52, 2004.

Index

"f" indicates material in figures. "t" indicates material in tables.

A

AaØ23 phage, 36
Abrasion, 54
Acanthamoeba, 101
Acetobacterium, 33t
Acholeplasma, 33t
Acids, organic, 48
Acinetobacter sp., 98, 102
Acinetobacter baumannii, 38, 96t
Acridine dye, 29
Actinobacillus spp., 25, 31t, 33–34, 33t
Actinobacillus actinomycetemcomitans, 26, 36
Actinomyces spp., 24
Actinomyces naeslundii, 27t
Actinomycetes, 111
Actinomycites, 98
Active exchange, 3
Active transport, 49
A-Dec ICX waterline tablets, 82–91, 86f, 87f
Adenosine triphosphate (ATP), 49
Adhesion; *See also* Coadhesion
    controlling, 54
    definition of, 50
    direct, 11, 12f
    establishment of, 50
    hydrodynamic forces and, 65, 100
    immune response and, 11, 65
    laminin and, 10t
    nonspecific, 50
    in oral cavity, 29f
    organic deposits and, 11f
    PIA, 63
    Si-Quat and, 56
    specific, 50
    studies of, 49

Adsorption, 55
AEGIS Antimicrobial, 47–48, 55
Aerobic bacteria, 28
Aerobic conditions, 3, 4
Aeromonas hydrophila, 96t
AFLP, 96t
Agglutinins, 24, 66; *See also*
    Hemagglutination
Air filters, 54
Air-fluid interface, 110, 140
Air-water syringe (AWS) lines, 83–85, 87
Albumin, 66
Alcaligenes, 33t
Alcohol, 17, 18t, 19, 84
Algae, 6, 54, 111; *See also specific types of*
Alimentary canal, 7, 26
Aluminum, 54
American Dental Association, 81–82, 90
American National Standards
    Institute (ANSI), 134–136
Aminoglycosides, 38
Ammonia, 123
Ammonium, 56, 102
Amoxicillin, 27
Amoxicillin-clavulanic acid, 27
Ampicillin, 30, 31t, 35
Amplicons, 117–120
Amplified fragment length
    polymorphism (AFLP), 96t
Anabolism, 3
Anaerobic bacteria, 30
Anaerobic conditions, 3, 4
Annular reactors, 124, 125
ANSI, 134–136
Antibody-specific probes, 64
Antimicrobials; *See also specific agents*

149

Printed and bound by CPI Group (UK) Ltd, Croydon, CR0 4YY

18/10/2024

01776267-0012